NISTIR 7941

Forensic Science Laboratories: Handbook for Facility Planning, Design, Construction, and Relocation

James Aguilar, AIA
Stantec Architecture, Inc.

Tom Barnes
Oregon State Police

Joseph Browne
Alison Kennedy
Romeo Miranda
Shannan Williams
Booz Allen Hamilton, Inc.

Yvette Burney
Scientific Investigation Division
Los Angeles Police Department

John Byrd
Central Identification Laboratory
Joint POW/MIA Accounting
Command

Bonnie Carver
Jim McClaren
Russell McElroy
McClaren, Wilson & Lawrie, Inc.

Adam Denmark
Michael Mount
SmithGroupJJR

Susan Halla
Lou Hartman
Kenneth Mohr
Crime Lab Design

Deborah Leben
United States Secret Service

Greg Matheson
Los Angeles Police Department

Steve Sigel
Virginia Department of Forensic
Science

Jennifer Smither
Science Applications International
Corporation

Melissa Taylor
Law Enforcement Standards Office
Office of Special Programs

Aliece Watts
Integrated Forensic Laboratories

June 2013

U.S. Department of Commerce
Cameron F. Kerry, Acting Secretary

National Institute of Standards and Technology
Patrick D. Gallagher, Under Secretary of Commerce for Standards and Technology and Director

Cover photographs of the Johnson County, Kansas, Sheriff's Office Criminalistics Laboratory taken by Joe Atchity and used with permission of Crime Lab Design, Inc.

CONTENTS

Contents

FIGURES

TABLES

In November 1996, the National Institute of Justice (NIJ), the National Institute of Standards and Technology's (NIST) Law Enforcement Standards Office (OLES), and the American Society of Crime Laboratory Directors held a joint workshop to develop guidelines for planning, designing, constructing, and moving into crime laboratories. The workshop's by-product, _Forensic Laboratories: Handbook for Facility Planning, Design, Construction, and Moving_, was published in April 1998 and was still in use up to the publication of this update. Over the 15 years since its original publication, however, significant changes have developed within the design and construction industry, specifically in regards to its focus on energy and sustainability. Additionally, dramatic advances in forensic science and research, and the resultant increased demand for forensic services have necessitated this first update to the 1998 handbook.

Today's capital investment decisions focus on the integration of new technologies, efficiencies, processes, and sustainable design/construction. The current economic climate demands an increased level of review and justification for each significant investment, especially in physical infrastructure and facilities.

In addition to the changes in the building industry and the economic environment, the increased demand for forensic science services has grown sharply in recent years. This increase is due in part to the increase in complex criminal activities that were not as prevalent when the handbook was first released. Cyber-crimes, increased terrorist activity, and advances in DNA analysis, in addition to the popularity of forensic science resulting from its glorification in television, have all driven an increased demand for forensic science services, and, therefore, an increase in the need for forensic science laboratory space.

BACKGROUND

The original _Forensic Laboratories: Handbook for Facility Planning, Design, Construction, and Moving_ was developed by a committee of 23 professionals from across the forensic science industry, including scientists, architects, engineers, laboratory directors, and representatives from state and local police laboratories. The group's unofficial charter called for the development of a resource for laboratory directors faced with building a new (or significantly renovating an existing) forensic science laboratory facility. The handbook was designed to help laboratory directors achieve three objectives related to their forensic science laboratory projects:

1. maximize organizational efficiency
2. ensure the economical expenditure of resources
3. develop a safe, secure, and well-designed facility

This first update to _Forensic Laboratories: Handbook for Facility Planning, Design, Construction, and Moving_ preserves the charter's original intent and builds upon its content to reflect the advances and current focus within the forensic science and facility construction industries.

PURPOSE

The information in this handbook provides laboratory directors, designers, consultants, and other stakeholders with guidance and tools to increase their knowledge of the project delivery process, their roles within that process, and their awareness of pertinent considerations. These considerations ensure that the end users receive an efficient, flexible, and functional facility that evolves with mission requirements. This handbook clarifies the laboratory director's role at each phase of the project life cycle, describes the impact of major decisions, and empowers users to ask the right questions at the right time. In addition, the handbook builds upon the original objectives of increased efficiency, fiscal stewardship, and sound facility design by doing the following:

- providing a more cohesive and structured document that follows a logical flow of information in a format standardized across chapters, thus facilitating document navigation and increasing document use
- incorporating updated policies, processes, and tools into the document and highlighting the relevance and anticipated benefits of each
- integrating considerations of statutory compliance requirements, as appropriate and applicable
- developing a document that can be leveraged by a broader audience to include laboratory directors, project managers, laboratory leadership, architects, engineers, and other project stakeholders as necessary

Each core phase of the facility delivery process leverages the collective insights, lessons learned, and best practices accrued by various forensic science laboratory facility projects already completed or currently underway.

SCOPE AND APPLICABILITY

This revised handbook is intended to serve as a resource for laboratory directors, designers, consultants, and other stakeholders involved in the construction or major renovation of forensic science laboratories. Each laboratory will have its own unique needs and requirements because of regional services and client base. Because many of the principles, processes, tools, resources, and considerations are applicable across a broad range of facility types, this document strives to highlight basic facility life cycle processes as well as those specific requirements that are unique to forensic science laboratories.

Although this handbook is intended to provide guidance and not to establish policy, the best practices and lessons learned outlined in this document should be strongly considered and integrated into new projects, where possible, to help ensure successful project execution.

TECHNICAL WORKING GROUP

The Forensic Science Laboratories Facilities Technical Working Group (TWG) was formed to support the preparation of this handbook. This group includes 16 professional forensic science planning experts with various backgrounds (laboratory directors, forensic science planners, architects, and engineers) from Federal, state, and local agencies as well as commercial architecture firms who participated in preparing this edition, *Forensic Science Laboratories: Handbook for Facility Planning, Design, Construction, and Relocation* (the "handbook"). In November 2011, the TWG (listed below) met for a working session to focus the document structure and to refine its content. The TWG members also reviewed draft versions of the document and provided additional content prior to final issuance.

TWG Members
James Aguilar, AIA, Senior Architect, Stantec Architecture, Inc.
Tom Barnes, Laboratory Director, Oregon State Police
Yvette Burney, Commanding Officer, Los Angeles Police Department, Scientific Investigation Division
John Byrd, Laboratory Director, Joint POW/MIA Accounting Command, Central Identification Laboratory
Bonnie Carver, Principal Architect, McClaren, Wilson & Lawrie, Inc.
Adam Denmark, Principal, Forensic Architect, SmithGroupJJR
Susan Halla, Project Leader, Crime Lab Design
Lou Hartman, Principal Engineer, Crime Lab Design
Deborah Leben, Laboratory Director, U.S. Secret Service
Greg Matheson, Laboratory Director, Los Angeles Police Department
Jim McClaren, Senior Principal Architect, McClaren, Wilson & Lawrie, Inc.

Russell McElroy, Senior Principal Architect, McClaren, Wilson & Lawrie, Inc.
Kenneth Mohr, Principal Planner, Crime Lab Design
Michael Mount, Forensic Architect, SmithGroupJJR
Steve Sigel, Deputy Director, Virginia Department of Forensic Science
Aliece Watts, Quality Director, Integrated Forensic Laboratories

Staff
Melissa Taylor, Project Leader, NIST, OLES
Shannan Williams, Associate, Booz Allen Hamilton, Inc.
Joseph Browne, Associate, Booz Allen Hamilton, Inc.
Alison Kennedy, Associate, Booz Allen Hamilton, Inc.
Romeo Miranda, Lead Associate, Booz Allen Hamilton, Inc.
Jennifer Smither, Technical Editor, Science Applications International Corporation

ABOUT THE SPONSORS

The National Institute of Justice (NIJ) is the research, development, and evaluation agency of the U.S. Department of Justice and is dedicated to researching crime control and justice issues. NIJ provides objective, independent, evidence-based knowledge and tools to meet the challenges of crime and justice. The Office of Investigative and Forensic Sciences is the Federal Government's lead agency for forensic science research and development as well as for the administration of programs that provide direct support to crime laboratories and law enforcement agencies to increase their capacity to process high-volume cases, to provide needed training in new technologies, and to provide support to reduce backlogs. Forensic science program areas include Research and Development in Basic and Applied Forensic Sciences, Coverdell Forensic Science Improvement Grants, DNA Backlog Reduction, Solving Cold Cases with DNA, Postconviction DNA Testing Assistance, National Missing and Unidentified Persons System, and Forensic Science Training Development and Delivery.

The National Institute of Standards and Technology's mission is to advance measurement science, standards, and technology. It accomplishes these actions for the forensic science community through the Law Enforcement Standards Office (OLES) Forensic Science Program. The OLES Forensic Science Program directs research efforts to develop performance standards, measurement tools, operating procedures, guidelines, and reports that will advance the field of forensic science. OLES also serves the broader public safety community through the promulgation of standards in protective systems; detection, enforcement, and inspection technologies; public safety communication; and counterterrorism and response technologies.

ACKNOWLEDGEMENTS

Forensic Science Laboratories Facilities Technical Working Group (TWG) gratefully acknowledges the following individuals for their contributions to the development and review of this handbook. Reviewers provided constructive suggestions but were not asked to approve or endorse any conclusions or recommendations in the draft handbook. Responsibility for the final content of this handbook rests with the members of the working group.

Ron Arndt, Lab Director, Weld County Police Department
Susan Ballou, Forensic Science Program Manager, NIST, OLES
Greg Czarnopy, Deputy Director, Bureau of Alcohol Tobacco, and Firearms (ATF)
Ramond DePriest, Forensic Lab Director, Metropolitan Police Department of Nashville and Davidson
Rita Dyas, Lab Director, Chandler Police Department
Steve Garrett, Forensic Services Director, Scottsdale Police Department

Steve Hackman, Principal, SmithGroup

Lindsey Horvat, Staff Assistant, Denver, CO Police Department

John Paul Jones, Working Group Program Manager for Forensic Sciences, NIST, OLES

Ken Konzak, Lab Director, California Department of Justice DNA Lab

Teresa "Terry" Long, Director, MD State Police Department

Pete Marone, Director of Technical Services, VA Department of Forensic Science

John Matthies, AIA, Senior Vice President, HDR

Mark Stolorow, Director, NIST, OLES

Ann Talbot, Assistant Director, Metropolitan Police Department of Nashville and Davidson

Richard Tontarski, Jr., Forensic Analysis Division, US Army Criminal Investigation Laboratory

William T. Vosburgh, Director, Forensic Science Division, Prince George's County Police Department

DISCLAIMER

It is NIST policy to use the International System of Units (metric units) in all its publications. In this report, however, information is presented in U.S. Customary Units (inch-pound), as this is the preferred system of units in the U.S. building industry.

1. DOCUMENT LAYOUT, PROCESS, AND ROLES

This chapter describes the document's layout, introduces the overall facility development and relocation process, defines the key roles of the process' participants, and describes the document's key format elements.

1.1 DOCUMENT LAYOUT

To simplify document navigation and to provide readers with a standardized approach for addressing each phase of the project delivery process, this updated handbook follows a common, logical, and standardized format that is consistent across the process chapters. Each section is composed of the following:

- project team roles and responsibilities within the particular process
- process diagrams and narrative descriptions
- tools to support each phase
- an actions checklist.

Many sections also include a list of frequently asked questions (FAQ).

 A blueprint icon highlights key deliverables produced in each of the phases of the project life cycle.

 The document also features quotes from various forensic science laboratory professionals. These quotes point out key insights these professionals had during the project experience.

> *"Quotes are placed in the document to draw attention to key insights professionals had during the project experience."*
>
> — *Process Participant*

1.2 PROCESS OVERVIEW

The life cycle of securing new laboratory space follows four key, primary phases – Planning, Design, Construction, and Relocation. This document models itself on these phases, one chapter for each phase.

These phases map out as illustrated on the following page in figure 1-1.

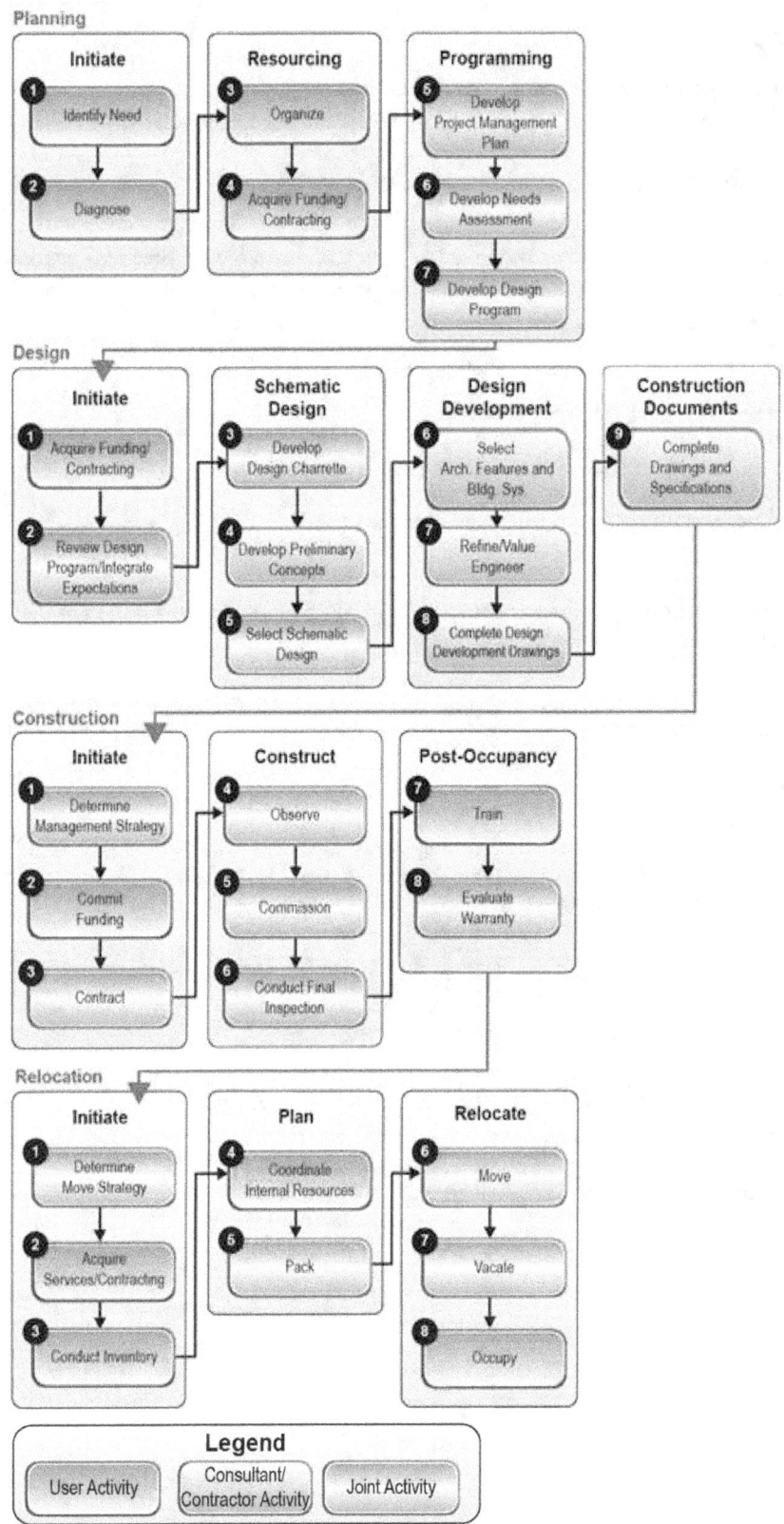

Figure 1-1: The forensic science laboratory Planning, Design, Construction, and Relocation process.

Document Layout, Process, and Roles

1.3   ROLES AND RESPONSIBILITIES

Facility design and construction projects rely on project teams with diverse memberships and skill sets for a successful outcome. A robust, involved, and integrated project team is essential to ensuring that new projects are appropriately coordinated, planned, and executed. This section introduces team members typical of those required to build a forensic

> "*The technical manager and laboratory Director need to define the project's expectations and inspire the staff to engage in the process as soon as the 'vision' begins to be 'planning.'*"
> —Laboratory Director

science laboratory and defines each member's responsibilities and the lines of communication among all team members.

Although specific team configuration is driven by the project's available resources (e.g., funding and organizational capabilities), the project's goals and constraints (e.g., deadline), and ownership hierarchy of the facility (e.g., facility owned or leased by project requestor), most projects will still include the four primary stakeholder groups listed below and included in figure 1-2. The amount of interaction between the parties will vary by phase. Maintaining the same team members, especially the laboratory director and technical manager, throughout the project's lifetime helps ensure delivery of the highest quality facility.

- **User Group:** The organization using the laboratory to perform its mission is referred to as the user group.

- **Property Owner:** The owner of the facility (if different than the facility user) is the property owner. Laboratories often lease space in a facility owned by a third party, such as another Government entity or a private owner.

- **Construction Team:** A group of contractors who will perform the physical construction of the facility is referred to as the construction team.

- **Design Team:** The design team consists of facilities design professionals who work with the user group to plan the laboratory and who prepare the construction documents needed

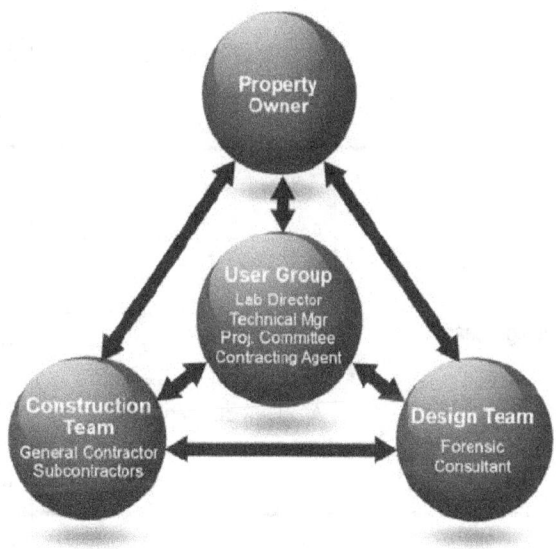

Figure 1-2: Project team.

to build the facility. The core design team includes, but is not limited to, forensic consultants; architects; interior designers; and mechanical, electrical, and plumbing engineers.

Table 1-1: Project Team Roles and Responsibilities Summary

Group	Role	Responsibilities
User (internal to laboratory)	Laboratory Director	Laboratory chief executive officer/project initiator and owner of all major decisions from the user's perspective. Navigates political and bureaucratic environments.
	Technical Manager	Internal team project manager. Ensures project effectiveness and progress on a day-to-day basis.
	Project Committees	Functional and technical capabilities experts. End users of laboratory space (functional) and/or technical staff members (e.g., information technology, security, etc.) who provide detailed subject-matter expertise focused on the specific functions, mission, or required technical requirements for the laboratory.
	Contracting Agent	Manages of all contracting requirements during project life cycle. Negotiates and executes agreements for all activities necessary to complete the laboratory space (e.g., consultant agreements or construction contracts).
Property Owner	Facilities Project Manager	Main point of contact for facility. If the facility is owned by entity other than user group, the facilities project manager represents the property owner's interests during project. Assists with coordination of design and construction with project team members.
Design Team	Forensic Science Consultant	Subject-matter expert on forensic science laboratory design. Specially qualified to prepare the Needs Assessment and Design Program. Offers value-added services throughout the balance of the project. Typically a multidiscipline team.
	Architect/Engineer (A/E) Team	Team of architects, mechanical engineers, security personnel, electrical engineers, commissioning agents, and other professionals (programmer, cost consultant, space planner, etc.) responsible for providing comprehensive A/E services throughout the Planning, Design, and Construction Phases. Multiple teams may be hired to complete different aspects of the projects.
Construction Team	General Contractor	Owner of construction contract. Responsible for delivery of constructed facility. Typically reports to property owner, laboratory director, and/or technical manager.
	Subcontractor	One who reports to only the general contractor. Performs specific construction tasks, such as designing and installing electrical systems, plumbing, and concrete work.

Various project delivery methods will impact the roles of each member noted in Table 1-1. Further description of the project team members' roles follow.

- **Laboratory Director:** The laboratory director is typically the project advocate and initiator. The director's primary engagement during the project is to drive any decision making that would impact the project's scope and funding during the Design Phase. The laboratory director may also choose to manage the day-to-day activities of designing and building the laboratory. Because of the heavy time commitment of project management, the laboratory director may consider designating a technical manager to perform the day-to-day functions. If this delegation is made, it is crucial to project success that the laboratory director maintains a high level of visibility during the entire life cycle of the project.

- **Technical Manager:** The technical manager is assigned to the project team by the laboratory director to represent the laboratory's interests and to act as a facilitator/coordinator during the project's life cycle.

  This team member represents the laboratory in the project's day-to-day tasks by overseeing, orchestrating, and coordinating the laboratory's project activities. The technical manager's primary responsibility is to make decisions on key elements of the project. Interaction with the project team includes responding to issues directly affecting the forensic science laboratory's

  > 66 *"Emphasize to your staff that if you don't get involved and speak up early in the project, it will likely be too late. Everyone has too much to do, but everyone needs to take the time if they want the best outcome."*
  > —*Laboratory Director*

  functional design requirements and facilitating the exchange of information between laboratory staff and any external team members (e.g., forensic science consultant) as necessary. To realize project success, the technical manager should be an able communicator who can effectively work with all team members. The technical manager should also possess working knowledge of the design and construction process, although the technical manager will most likely not possess formal construction management training. The ideal candidate is a scientist with construction knowledge. Performance of the technical manager role can be a full-time job and, depending on the project, will most likely require that the technical manager temporarily set aside some, if not all, other job responsibilities. Based on expertise and time requirements, the technical manager's role may be supported by an owner's representative.

- **Project Committees:** The two project committees, the functional committee and the technical committee, provide review and input to ensure the effectiveness of the constructed laboratory. Each includes internal stakeholders and experts who are the end users of the laboratory space. Their roles during the project process are to field and answer questions from the design team and to review and comment on major milestone design submissions. These activities focus on ensuring the desired functionality of the laboratory. The functional committee assists with ensuring the effectiveness of the laboratory's mission-essential functions. This committee is composed of scientists and, ideally, one staff member from each laboratory section. The technical committee focuses on operational effectiveness of the laboratory and is composed of individuals involved in information technology (IT) and security. Both committees are assembled by and report to the technical manager.

- **Contracting Agent:** The contracting agent is responsible for administering all contract documentation on behalf of the laboratory. The contracting agent provides legal advice and subject-matter expertise on the acquisition process to assist the laboratory leadership with decision making. To ensure success, the technical manager must understand the contracting agent's responsibilities. Both must work together when administering acquisitions. The contracting agent's main point of contact is the contracting officer within the agent's organization, which can vary depending on the laboratory. Most construction contracts are administered through a facilities management/planning or contracting office. Some organizations do not have internal contracting teams. In this case, leadership must secure contracting support through outside organizations or engage a construction management firm outside the construction agreement to administer acquisition.

- **Facilities Project Manager:** When the laboratory facility is not owned by the user group, a facility project manager is assigned by the property owner to participate in the project. This individual assists the Design and Construction Phases by providing building information (e.g., base drawings) and by performing any activities as required by the lease agreement. The ideal candidate for this role is the organization's current facility manager.

- **Forensic Science Consultant:** The forensic science consultant prepares most deliverables during the Planning Phase, including the Needs Assessment and Design Program. The forensic science consultant interacts with the technical manager and project committees throughout the Planning Phase to ensure that the project requirements are thoroughly captured, documented, and integrated into the Design Program. The forensic science consultant submits major deliverables within the Planning Phase. The forensic science consultant is typically retained during the Design and Construction Phases to ensure that design requirements are appropriately integrated into the design solution. Responsibilities may vary from an oversight role to preparation of deliverables specifically related to the forensic laboratory design. If the property owner or laboratory director hires the forensic science consultant, this consultant reports directly to the technical manager. However, this consultant is typically part of the A/E team and, in this case, reports directly to the principal in charge of the A/E team.

- **Architect/Engineer Team:** The A/E team prepares all other deliverables during the design process. The A/E team possesses the subject-matter expertise to develop a constructible solution for the laboratory and interacts with the user group throughout the design process to ensure the final design effectively deliver's mission requirements. The A/E team submits major deliverables, as described in the Design process section and finalizes construction documents that will be used to build the project. The A/E team is hired at the project inception and is retained through the Construction Phase in an oversight (administrative) role to ensure that construction meets the standards defined in the drawings and to address unexpected site conditions. The A/E team reports directly to the technical manager.

- **General Contractor:** The general contractor is responsible for the actual construction onsite. The general contractor's staff may perform some or all of the construction or may use subcontractors.

- **Subcontractor:** Subcontractors are engaged by the general contractor to perform pieces of the construction that the general contractor does not possess the ability or capacity to perform. Subcontractors often include specialty contractors, such as plumbers or electricians. Subcontractors' communication is routed through the general contractor.

Document Layout, Process, and Roles

As with any new project or major renovation/repair project, proper planning is the first step in ensuring project success. Issues during the Design, Construction, and Relocation Phases can be greatly reduced through careful and deliberate planning at the start of the project life cycle. This includes ensuring that the right stakeholders are integrated into the project delivery process as early as possible and that the project team has access to subject-matter experts (SMEs) to help guide them through the complicated and time-consuming process of planning a new facility or major renovation. An integrated project team involved throughout the process is critical to ensuring project success.

A deliberate planning process ensures (1) that project owners are designing and building the right facility for their particular needs now and into the foreseeable future and (2) that capital investment projects not only meet the mission needs of the requesting organization but also integrate and support overall agency or enterprise goals and objectives related to mission effectiveness.

2.1   PLANNING PHASE ROLES AND RESPONSIBILITIES

During the Planning Phase, active project team members should be the individuals identified in table 2-1. Additional project team members may be required depending on the project scale, organizational construct, or special requirements.

Table 2-1: Project Team Activities during Planning Phase

Group	Role	Planning Phase Responsibilities
User (internal to laboratory)	Laboratory Director	Leads project visioning. Selects team. Ensures completion of deliverables. Builds consensus for project advocacy.
	Technical Manager	Manages day-to-day activities. Takes lead on development of Project Management Plan. Assists with team assembly.
	Project Committees	Assist with development of Needs Assessment and Design Program.
	Contracting Agent	Assists in contracting consulting services. Assists planning and decision making by providing guidance on contracting options for entire project life cycle.
Property Owner	Facilities Project Manager	Maintains project awareness. Provides input on how facility constraints impact Design Program options.
Design Team	Forensic Science Consultant	Leads the development of the Needs Assessment and Design Program. Key external resource.
	A/E Team	*Not typically engaged.*
Construction Team	*General Contractor*	*Not typically engaged.*
	Subcontractor	*Does not participate.*

2.2   THE PLANNING PROCESS

The process of planning for a new facility or major renovation is composed of three distinct stages in which the project owner and related staff members must be intimately involved. The process depicted in figure 2-1 summarizes the primary stages of the Planning Phase: initiating, resourcing, and programming. Each of these stages is described in detail in the sections that follow.

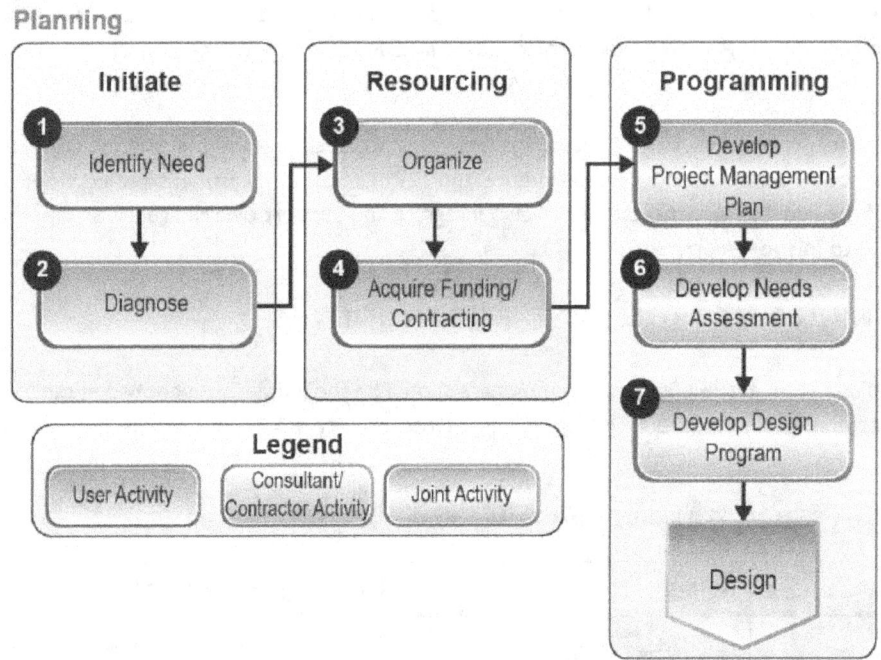

Figure 2-1: Forensic science laboratory project Planning Phase.

2.2.1   INITIATE

Every new project or major renovation request starts with an identified need, which can be driven by a number of considerations, including, but not limited to, advances in technology or methodologies, lack of existing space/capacity, growth in staff or scientific requirements, increased demand for forensic science services, or the need to update facilities and major infrastructure systems because they have outlived their design life. The initiation of a new project begins with identifying the need and then diagnosing existing operations, capabilities, and requirements to start the process of *defining* the need. Specific steps within the initiate stage of the Planning Phase are illustrated in figure 2-2.

❶ **Identify Need:** This step establishes a decision to pursue a major laboratory infrastructure project. The need for a project can be received from or conceived by any number of sources, such as research staff members, laboratory directors, operations and maintenance (O&M) staff members, or headquarters. As noted above, the need can also be driven by any number of considerations, such as mission growth, increased demand for forensic science services, and inadequate space.

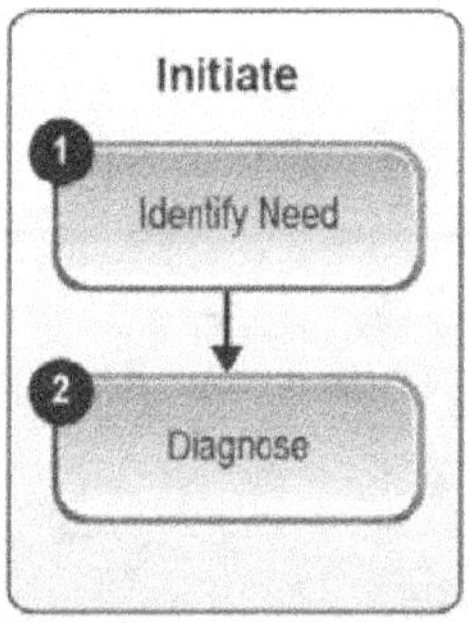

Figure 2-2: Initiate stage of the Planning Phase.

② **Diagnose:** This step commences the process of defining the scope of the laboratory infrastructure project. Once the need for a new facility project is preliminarily determined, the agency performs a Self-Evaluation (deliverable) to define the mission requirements that must be addressed by a new facility or existing facility renovation. The evaluation should highlight current, past, and future (projected) mission requirements, as well as organizational requirements. These mission and organizational requirements should be developed collaboratively among laboratory staff, management, headquarters components, etc., to ensure that all potential stakeholders have the opportunity to define the need.

The activities outlined in the initiate stage of the Planning Phase are typically performed by laboratory staff members, specifically the laboratory director and/or his or her designated technical manager.

 Initiate Stage Best Practices

- Use a self-evaluation survey or questionnaire to identify deficiencies in the current laboratory.
- Predict growth trends and consider other external influences.
- Look for partners to strengthen the project justification.
- Bolster project justifications by leveraging data from growth trends and external influences, such as court-mandated changes, crime data, demographics, case loads, and economic trends. Fully develop these data points during the Needs Assessment process.
- Rethink existing laboratory processes, as they may not be the ideal benchmarks for developing project justifications. If possible, take tours of newer facilities.

FAQ

Q: *What questions should I (laboratory director) ask to support the advocacy/justification for my proposed project?*
A: Ask the following questions:
- Is the project essential to ensuring the safety and security of both assigned personnel and/or facility operations?
- Is the project required to meet current/new/emerging regulatory compliance requirements, or will it improve the margin of compliance?
- Is the project required to improve the integrity/handling/preservation of evidence?
- Will the project improve mission performance, evidence analysis, and/or other critical operational requirements?
- Is the project required to avoid a highly probable critical system(s) failure and/or provide mission redundancy to maintain mission reliability?
- Have the existing facility's major infrastructure systems reached/exceeded their estimated design life/system capacity?
- Is the project required to address a change/growth to the current mission and associated staffing?
- Is the project required to provide a new capability/technology/process?
- Will the project contribute to meeting sustainability and/or energy management goals/objectives?

2.2.2 RESOURCING

The resourcing stage of the Planning Phase focuses on assembling the project team to begin formal planning activities. As shown in figure 2-3, this stage includes organizing initial team members, securing funds, and entering agreements for external consultant services.

❸ Organize: After validating the need for a new facility project, the laboratory director should appoint a technical manager to oversee the overall execution and oversight of the project activities. The technical manager starts to develop a project team with the required competencies to ensure a successful project and to establish initial planning activities, such as developing a baseline budget, scope, and schedule. The technical manager and project team may also review strategic plans, master plans, market analyses, etc., to ensure that the proposed project is compatible with agency long-range goals and/or local community plans (e.g., another local agency may also be planning a similar facility that may present partnering opportunities).

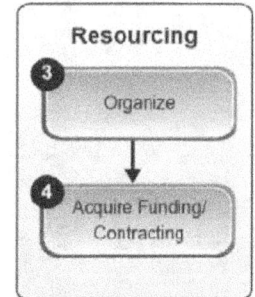

Figure 2-3: Resourcing stage of the Planning Phase.

At this time, the technical manager also typically begins to initially conceptualize the project management plan (PMP), which outlines the strategy for planning, designing, constructing, commissioning, and relocating into the facility. The PMP is a dynamic document that is continually updated and refined throughout the project life cycle.

❹ Acquire Funding/Contracting: Concurrent with the organize step, the technical manager, with appropriate support from the project team, identifies the types of external consultants necessary to build out the project team.

> ❝ *"I can't emphasize enough the direct relationship of using the same forensic consultant throughout the process to owner satisfaction."*
> —*Laboratory Director*

The technical manager identifies the external capabilities required to guide, execute, and/or support the development of planning deliverables, including the Needs Assessment and Design Program. These resources are to be familiar with the specifics of forensic science laboratory design.

The technical manager should first check within the organization's jurisdiction and within sister agencies for internal planning resources. If sufficient forensic science laboratory planning expertise is resident within the agency, the technical manager may elect to develop the Needs Assessment and Design Program with in-house resources and forgo external consultants.

Once the external capabilities have been identified, the technical manager seeks funding to contract for forensic science consultant services. In this regard, the technical manager submits the baseline budget, scope, and schedule (developed in the organize step) with the initial project justification (developed in the diagnose step) to the agency-specific budget and finance group for funding evaluation.

Upon approval of funding, the technical manager, along with the project team, develops a statement of work (SOW) and a request for proposals (RFP) for consultant planning services (see sample SOW/RFP at http://www.nist.gov/oles) The SOW and RFP should integrate considerations of health and safety (such as liability issues), building codes (such as fire and life

> ❝ *"A political champion to authorize the massive expenditure will help smooth the road of potential government logjams."*
> — *Laboratory Director*

safety), new technologies (including existing and projected new technologies), crime trends, and existing constraints (such as finances, personnel, and location).

The resourcing stage ends with the selection of a forensic science consultant and the issuance of a notice to proceed (NTP).

Resourcing Stage Best Practices

- Maintain a stable team for project continuity. As the project progresses, the roles and responsibilities of the project team evolves, so stability is important.

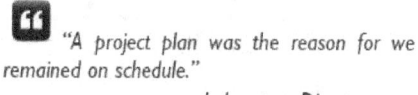

> 66 *"A project plan was the reason for we remained on schedule."*
> —*Laboratory Director*

- Rely on SMEs to help plan projects. Specifically, use forensic science consultants to advise and help manage the Planning Phase. Their expertise ensures a properly scoped, budgeted, and planned project.

- Check consultants' references and past performance; a qualifications-based selection criterion is essential.

- Select a forensic science consultant with A/E capability so the consultant can be used during the Design Phase.

- Know your advocates, such as parent organization and interested elected officials. They may be able to assist in moving the project forward.

- Develop a working knowledge of the requesting agency's specific budget cycle, processes, and submittal requirements to ensure that the appropriate consideration is given to the requested project.

- Emphasize the importance of the project team's involvement throughout the process, especially regarding anticipated daily facility operation.

- Document the project's organizational chart and share it with the project team.

2.2.3 PROGRAMMING

The programming stage, outlined in figure 2-4, requires the project team to develop the specific project requirements, strategy, and management structure through several deliverables, including a PMP, a Needs Assessment, and a Design Program. This process defines the laboratory's program.

 ❺ **Develop Project Management Plan:** The technical manager, with the forensic science consultant, prepares the project management plan. The PMP baselines the execution approach and controls to be used throughout the project life cycle. This plan commences with identifying the project deliverables and defining the work breakdown structure (WBS) of activities required to complete the deliverables. Resources and time estimates to complete these activities are then translated into the project schedule. Once the schedule is prepared, the project can start with leadership possessing a detailed awareness of the most efficient budget and time management tools for successful implementation.

**Figure 2-4:
Programming stage of
the Planning Phase.**

The PMP also formalizes the project team and clearly outlines each member's roles and responsibilities. The most effective project teams include both functional (e.g., laboratory staff members) and technical representatives (e.g., information technology, security). At minimum, the team should include the members listed in table 2-2. A project kickoff meeting between the user group and the forensic science consultant reviews the PMP in detail. Typically, the consultant coordinates the kickoff meeting with the project owner's technical manager.

Table 2-2: Project Team Members

User/Agency Representatives	External Representatives
Laboratory Staff	Architect/Engineer
Facilities/Real Estate	Programmer
Public Works	Cost Consultant
Budget/Finance	Space Planner
Contracts/Contracting Officer/Contracting Officer's Representative	Others as needed (e.g., environmental or traffic engineers/consultants)
Information Management/Information Technology	Forensic Science Consultant
Security/Safety/Environmental	Building Owner (if different from agency)

❻ Develop Needs Assessment: The Needs Assessment is an essential planning tool that is typically developed by independent professionals (e.g., forensic science consultants). This document considers research, user, and facility needs; assesses the existing facility condition, limitations, and challenges; defines all space requirements; and refines preliminary opinion of construction cost. The Needs Assessment also integrates considerations of external influences, including building codes; Federal, state, and local requirements; and sustainability and green building goals and objectives.

Although the forensic science consultant typically leads the Needs Assessment development process, the technical manager is integral to the process and must be fully involved to ensure that the consultant has access to the required information, staff resources, facilities, processes, and other information needed for a comprehensive and accurate assessment of facility needs. With the required access to the appropriate user resources, the forensic science consultant uses a three-step process for developing the Needs Assessment as seen in figure 2-5.

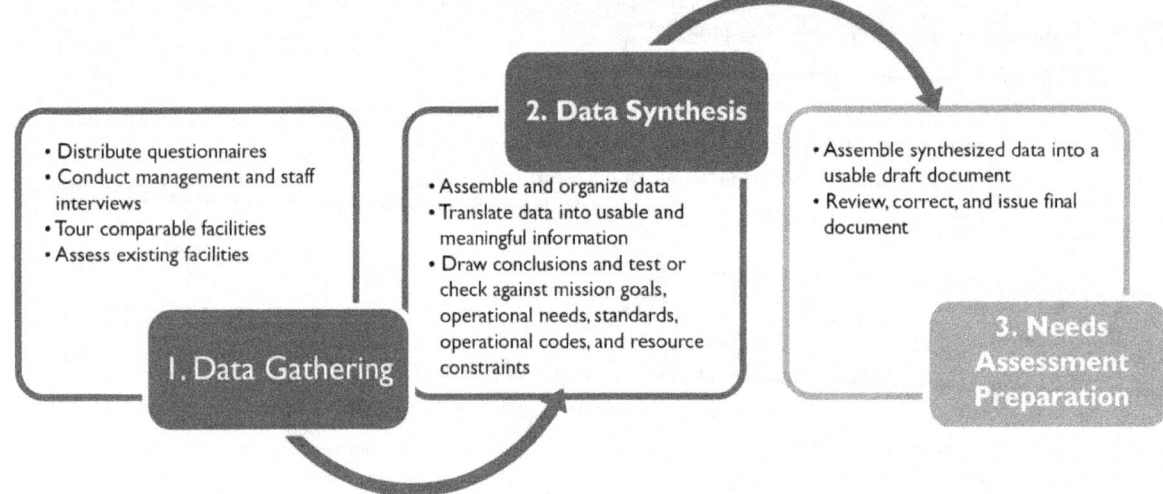

Figure 2-5: Needs assessment process.

The Needs Assessment answers the following key questions: How big (size)? How much (cost)? Where (site evaluation and selection)? and When (schedule)? It may also include an assessment of the condition and deficiencies of the existing facilities to help justify the need for a new laboratory. The completed Needs Assessment may be used by the laboratory director as a tool to request funding for a new laboratory. An outline of the various components of the Needs Assessment is provided in appendix E. A sample completed Needs Assessment is at http://www.nist.gov/oles/.

 ❼ Develop Design Program: The Design Program complements and further develops the Needs Assessment by translating requirements identified in the Needs Assessment into data that are used by architects and engineers to design the facility. Frequently, the Design Program and the Needs Assessment are combined into a single document.

The Design Program is developed by the forensic science consultant, but, like the Needs Assessment, it requires considerable input from the user group, including laboratory staff members, management, and other potentially affected departments (e.g., project team). The Design Program outlines the concepts that satisfy the baseline budget, scope, schedule, sustainability, security, and safety requirements outlined in the Needs Assessment. As appropriate, strategies are also determined for Building Information Modeling (BIM), commissioning and Leadership in Energy and Environmental Design (LEED) certification.

At the conclusion of the Design Program step, a single concept is selected to proceed to the Design Phase, and the forensic science consultant, with the technical manager, develops a detailed schedule and cost estimate based upon the selected concept. The user group may also choose to conduct a value engineering analysis, either in-house or through another consultant, to ensure that the selected concept provides the best value to the user group while also meeting project and agency goals.

Although the majority of activities and deliverables (e.g., Needs Assessment and Design Program) in the Planning Phase are led by the forensic science planning consultant, the user group plays an integral part at each step to ensure that user-specific requirements and processes are captured, integrated, and appropriately addressed during programming is critical. A sample completed programming document is available at http://nist.gov/oles.

 Programming Stage Best Practices

- Because laboratory staff members (users, management, and others) are critical to the Needs Assessment process, ensure that they are available to provide input into research needs, special spatial requirements, required adjacencies, technology requirements, and other process-related considerations that will direct Design Program development.
- Ensure that a Needs Assessment is done as early as possible by a qualified professional to establish the project's definition. This enables effective budgeting as early in the process as possible.
- Ensure non-construction-related costs are integrated into the total funding picture. See appendix D: Cost Considerations for a list of non-construction-related costs to consider.
- As a laboratory director, do not respond to requests for project specifics (e.g., from the media or superiors) until the Needs Assessment is complete (including scope and cost).
- Carefully plan storage space into new projects and build strong justifications to warrant the space. Without this, the storage space becomes occupied space as the scientific need grows.
- Base the project's requirements upon baseline and future growth milestones. The Needs Assessment should incorporate future growth milestones.
- Discuss facility projects with colleagues, other forensic science service agencies, and local area partners for A/E team recommendations based upon past experience and performance on other forensic science laboratory projects.

- Keep a forensic science consultant on the A/E team during the Design and Construction Phases if possible.

FAQ

Q: *What can I expect as I start the Needs Assessment and Design Program processes?*
A: The Needs Assessment and Design Program must do the following:

- Demand full user/client participation.
- Show that the final product is not just a customer "wish list"; Provide justifications for the proposed size and cost.
- Result in resolving issues related to expansion, splitting operations, renovations/expansions, or building a new facility.
- Include detailed information on the size and proposed costs of the proposed facility and the building's location.

2.3 PLANNING PROCESSES AND TOOLS

Although the Planning Phase is quite involved and relies heavily on user group engagement on the various activities described above, the user group has access to a number of tools and resources to provide further assistance. This section discusses some of these tools. Also, appendix A: Standards, Regulations, and Other Guidance provides additional planning resources.

2.3.1 SPACE-TO-STAFF RATIOS

The size of the laboratory determines the building cost. Laboratory size can be estimated early in the planning process using a space-to-staff ratio rule of thumb. Although space standards vary widely by organization, using a range of 700 to 1,000 square feet per staff member offers a quick snapshot of the laboratory's potential size. As the size of the facility increases, economies of space reduce the space-to-staff ratio. For example, regardless of the size of the crime laboratory, it requires only one firing range, one reception/waiting room, and one space for standards and references. Using building categories of small, medium, large, and very large, a project team can use the ratios outlined in table 2-3. These gross square foot (GSF) figures account for a pro-rated portion of support spaces, such as horizontal and vertical circulation, utility rooms, mechanical rooms, electrical rooms, and shared laboratory space. This table does not indicate the amount of space allocated for individual use.

Table 2-3: Average Space-to-Staff Ratios for Laboratories

Laboratory Category	Staff Size	Total Laboratory Size	Space per Staff Member
Small	Up to 30	Less than 30,0000 GSF	930 to 1,000 GSF per staff member
Medium	30 to 70	30,000 to 60,000 GSF	860 to 930 GSF per staff member
Large	70 to 110	60,000 to 90,000 GSF	790 to 860 GSF per staff member
Very Large	Over 110	More than 90,000 GSF	720 to 790 GSF per staff member

Many space factors associated with forensic science laboratories may skew these ratios. For example, a crime laboratory might include large unoccupied spaces (property and evidence storage or a large training center), which will drive the ratio down. In addition, these ratios are not applicable to forensic science laboratories in foreign countries, as many local customs and traditions will influence the space-to-staff ratios.

Again, these numbers represent only a "rule of thumb," whose use is limited to generating a rough, initial estimate of a laboratory's size and are not recommended for establishing a project construction budget. These numbers are not a substitute for a comprehensive Needs Assessment prepared by an A/E team with forensic science laboratory planning and design experience.

2.3.2 ALTERNATIVE FUNDING OPTIONS

One of the primary challenges laboratory directors face when seeking approval to move forward with a new or renovated laboratory project is justifying the expense against increasingly shrinking capital budgets and competing operational requirements. Often, physical infrastructure projects undergo the most scrutiny and are usually the first requirements to be cut from shrinking budgets, as many cash-strapped Federal, state, and local agencies make do with existing facilities. One strategy is to secure funding from sources independent of annually recurring organizational budgets. Options to consider include general obligation bonds, partnerships (e.g., academic institution, other government agency, or public-private partnership), tax riders, and/or lease purchase options.

> *"When community acceptance is important, an outreach program informing residents of what is coming will help keep the project moving smoothly."*
>
> — *Laboratory Director*

2.3.3 PLANNING PHASE CHECKLIST

The following checklist, in table 2-4, identifies the activities that must be completed as part of performing the Planning Phase. Checklists for all phases are in appendix F: Key Activity Checklists.

Table 2-4: Planning Phase Checklist
(Note: bolded items are Phase deliverables)

	Process Step	Activity
☐	2	Mission Requirements Identified in **Self Evaluation**
☐	3	Technical Manager Appointed
☐	3	Project Team Assembled
☐	3	**Project Management Plan** Initiated
☐	3	Baseline Project Budget Developed
☐	4	External Consultant Identified
☐	4	Consultant Funding Acquired
☐	4	Consultant SOW/RFP Defined
☐	4	Consultant Contracted
☐	5	Project Team Completed
☐	5	Project Management Plan Completed
☐	6	**Needs Assessment** Prepared
☐	7	**Design Program** Prepared

There is not a universal plan for a forensic science laboratory design. Design is driven by types of evidence, laboratory processes, equipment, and future needs. These considerations are specific to each individual laboratory. Functional requirements of specific scientific disciplines, equipment, and instrumentation affect the space, dimension, and adjacency requirements of the final design. For example, some laboratories must consider spaces for hazardous materials handling and preservation of evidence. Each laboratory must also consider its unique future requirements and opportunities. A forensic science laboratory designed with the flexibility to change along with the needs of its occupants, technology, or scientific methodologies multiplies the efforts of the staff members working within it.

Appendix B: Laboratory Design Considerations provides descriptions of the major elements of a laboratory design (e.g., plumbing and laboratory design standards) for reference during the Design Phase.

3.1   DESIGN PHASE ROLES AND RESPONSIBILITIES

During the Design Phase, the A/E team takes the lead in preparing the deliverables. However, direct involvement by the user results in a higher quality final design. Table 3.1 highlights some of the project team's activities specific to the Design Phase.

Table 3-1: Project Team Activities during Design Phase

Group	Role	Responsibilities
User (internal to laboratory)	Laboratory Director	Remains engaged in the decision making process. Participates in each design session.
	Technical Manager	Manages day-to-day activities of process. Ensures the heavy participation of the Project Teams.
	Project Committee	Participates extensively in design decision making, especially during Design Development. (Note: Design Phase offers the greatest positive project impact for the project team.)
	Contracting Agent	Assists in contracting consulting services.
Property Owner	Facilities Project Manager	Participates in all design sessions to define the extent to which construction can impact existing structures and configuration for ongoing facility maintenance.
Design Team	Forensic Science Consultant	Maintains participation in Design Phase to ensure effective technical design.
	A/E Team	Acts as the lead consultant in the Design Phase. Prepares all major deliverables.
Construction Team	General Contractor	May occasionally help validate proposed design concepts within construction budget. Involvement contingent on project delivery method (e.g., design-bid-build)
	Subcontractor	*Does not participate.*

Design

3.2   THE DESIGN PROCESS

The Design Phase offers an iterative decision making approach for developing a buildable solution. This phase, seen in figure 3-1, requires participants to make the right decision at the right time by validating function, cost, and construction detailing, in that order.

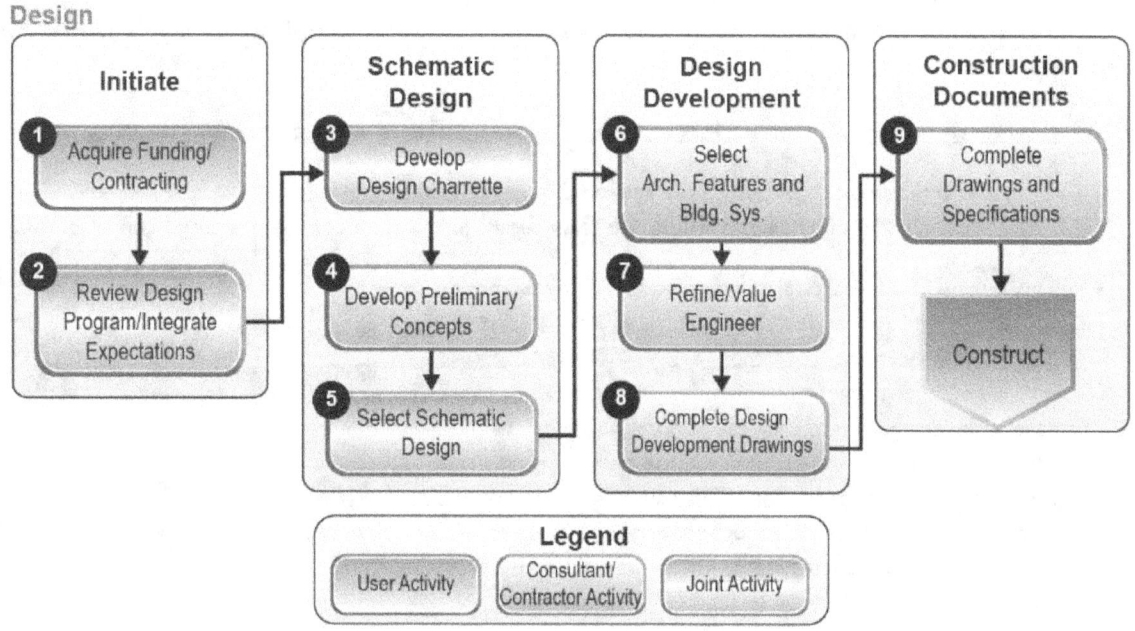

Figure 3-1: Forensic science laboratory project Design Phase.

3.2.1   INITIATE

The initiate stage includes contracting an A/E team and then introducing the A/E team to the project, as shown in figure 3-2. This offers an opportunity for additional refinement of the project program.

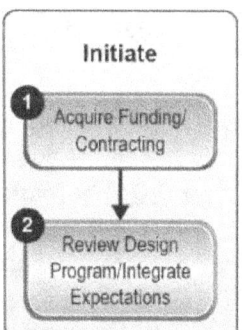

Figure 3-2: Initiate stage of the Design Phase.

❶ **Acquire Funding/Contracting:** The user group procures the services of an A/E team to prepare the design documentation required to construct the laboratory. The technical manager identifies which external capabilities are required to guide and prepare the development of design deliverables. Once the external capabilities have been identified, the technical manager commences the process of acquiring funding to contract for A/E team services. The technical manager submits the baseline budget, scope, and schedule information to his or her agency-specific budget and finance group for funding evaluation. The technical manager first checks the laboratory organization's jurisdiction and sister agencies for internal design resources. Design resources considered for contracting will possess experience with the specifics of forensic science laboratory design.

Upon approval of funding, the technical manager, along with the project team, develops an SOW and an RFP for the A/E team's services (see sample SOW/RFP at http://www.nist.gov/oles/). After the A/E team is selected and awarded, the technical manager gives the A/E team an NTP in preparing the laboratory's construction documentation.

Although unlikely, if sufficient forensic science laboratory design expertise is resident within the agency, the technical manager may elect to develop construction documentation with in-house resources and to forgo the need to contract with external consultants to provide these services.

> *"Make sure your A/E team has relevant experience and make sure to get and check references of your bidder. Ask other users how the A/Es were to work with."*
> —*Laboratory User*

2 **Review Design Program and Integrate Expectations:** There are two goals for this step:

1. introduce the A/E team to the desired Design Program;
2. secure initial feedback from the A/E team regarding program completeness and constructability.

This review may be particularly valuable as frequently months or years may pass between the completion of the program and commencement of the design. The A/E team recommends adjustments/refinements to the project scope. The technical manager confers with the project team on recommendations. The technical manager approves the recommendation if appropriate. This step may include several team meetings for the introduction of scope and review of recommendations. During this step, the project team must comprehensively communicate the project's Design Program and intent and capture feedback from the A/E team.

> *"Face-to-face meetings at initial information gathering and review of major milestones with entire team ensure effective communication with the A/E."*
> — *Laboratory Director*

Once feedback is obtained, the A/E team formalizes both the adjustments to the Design Program given the approvals provided by the technical manager and the goals and standards that the project aims to achieve (e.g., LEED certification level). During this step, the project team must review the A/E team's recommendations and incorporate desired adjustments to project Design Program.

Initiate Stage Best Practices

- Be prepared to capture additional value for the project through input from the A/E team because they provide key insights into the project's program based upon its depth of experience in the design and construction environment.
- Do not constrain the design to existing functionality. Evolve the functional capability during the design. Do not design for how the process works currently but rather for how it should be working. Current processes are often hampered by the conditions of the existing facility.
- Design to accommodate the desired laboratory culture (e.g., collaboration with open seating or privacy with separate offices).
- Ensure communication with the A/E team during the Design Phase. Ensure that all voices are heard and that all participants ask questions.
- Include a checklist on a kickoff meeting agenda, including nomenclature, process description, and plan symbols.
- Ensure that the same forensic science consultants (if possible, the same individuals) are used continuously throughout the project, from planning through construction.

FAQ

Q: *What is the expected level of involvement during the Design Phase?*
A: The level of involvement varies throughout the process and among roles. Laboratory directors can expect to spend about 25 percent to 33 percent of their time on design, whereas technical managers are involved 100 percent of the time. Other user group staff can expect to spend dozens of hours contributing to the design.

3.2.2 SCHEMATIC DESIGN

As shown in figure 3-3, the Schematic Design translates the project scope into conceptual design drawings possessing all the functionality required of the project. Clear decisions on the site design are completed, resulting in initial floor plans with facility elevations. Basis of design specifications are prepared, and all major components and equipment are defined.

❸ Develop Design Charrette: The design charrette is an intensive, collaborative session or series of sessions where project members validate the functional space requirements of the project scope against space available on the site. This step offers the opportunity to resolve functional adjacencies. The project team may also provide the A/E team with additional details on exactly how they desire the space to work. This is a whole-team exercise, typically completed in a working meeting format (requiring half a day to multiple days, depending on project size). Bubble diagrams showing the practical layout of the project as well as some initial indications of systems and materials to be used onsite are typically provided in the charrette. The process may prepare functional inventories (e.g., equipment and chemicals used in the laboratory) that continue to be developed through the entire design process. The design charrette process also identifies key problem issues and additional information needed by the A/E team to begin the design process. These issues are tracked as action items for the technical manager and project team to provide a response. The project team must participate in developing the design charrette and must finalize communication on defining goals for the project.

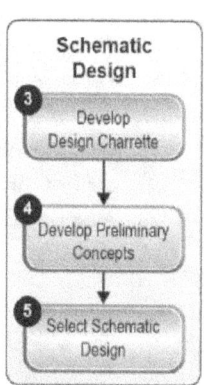

Figure 3-3: Schematic design stage of the Design Phase.

❹ Develop Preliminary Concepts: The goal of this step is to develop several plans of how the project's functions can work on the site. The A/E team generates these concepts based upon the Design Program and the outcomes from the design charrette. The plans will be completed to a detailed level and identify basic information on major systems. During this step, the project team answers all queries from the A/E team.

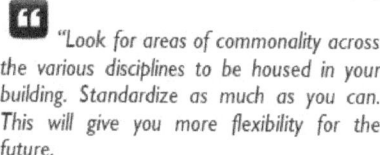

> "*Look for areas of commonality across the various disciplines to be housed in your building. Standardize as much as you can. This will give you more flexibility for the future.*
>
> —Laboratory User"

❺ Select Schematic Design: During this step, the technical manager reviews concepts to ensure that site plans possess the full functionality desired for the project. These plans constitute the Schematic Design (deliverable) and include a construction cost estimate. Functionality is locked in at this point to enable the design to focus on the next layer of decisions. The project team approves the design and cost estimate, changes each as necessary to match the project's intent and to fit within resources, and engages additional support from SMEs to perform tasks as needed.

Schematic Design Stage Best Practices

- Incorporate sustainable and energy-efficient design concepts into the schematic design stage. Sustainable design seeks to reduce negative impacts on the environment and the health and comfort of building occupants, thereby improving building performance. The basic objectives of sustainability are to reduce the consumption of nonrenewable resources; to minimize waste; and to create healthy, productive environments. Under Executive Order 13423, *Strengthening Federal Environmental, Energy, and Transportation Management*, the U. S. Government instructs Federal agencies to conduct environmental, transportation, and energy-related activities under the law in support of their respective missions in an environmentally, economically, and fiscally sound, integrated, continuously improving, efficient, and sustainable manner.

- If using a construction manager (CM) or an independent project manager, get this person involved early in the design process.

- Fully engage the user's SMEs during major milestone reviews to ensure the elements within their expertise remain on track.

- Perform a life cycle cost analysis to reduce overall facility cost with an awareness of the expected operating costs. The design and construction of the facility are a fraction of the cost of operation.

- Ensure that laboratory directors remain highly engaged in the schematic design stage. Participate in each step of the design process and in every design session. Delegate participation to staff members as needed.

- Maintain an aggressive approach to receiving and processing documents for review. Delays in approval will marginalize the A/E team's efforts and quality of the final product and can impact costs through additional services.

3.2.3   DESIGN DEVELOPMENT

Design development ultimately provides a plan identifying a detailed solution (e.g., specific materials and systems) that can be built within budget. At this stage, seen in figure 3-5, documents, which include both drawings and outline specifications, are generated. Systems and equipment requirements continue to be developed. Appendix B: Laboratory Design Considerations provides a broad checklist of considerations that laboratory directors and their staff members should be cognizant of during the Design Phase. The laboratory's project committees have their greatest involvement during this stage. Also during this stage, the greatest tension for decision making exists because the project team has to make numerous decisions between competing interests. A high level of user participation at this stage is essential to limiting expensive changes in future project phases. As depicted in figure 3-4 below, the cost of integrating changes into the design grows exponentially as the Design Phase progresses.

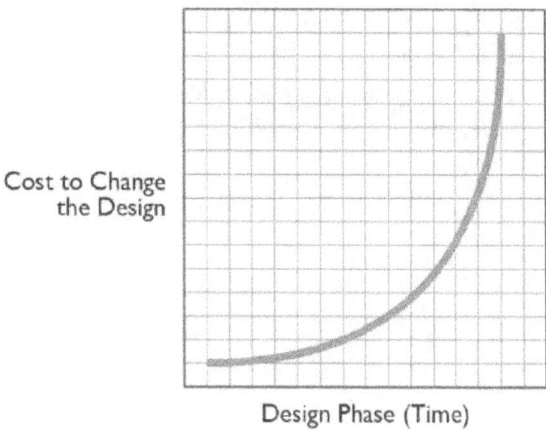

Figure 3-4: Increasing cost of changes as a function of time.

⑥ <u>**Select Architectural Features and Building Systems:**</u> The A/E team works with the project team to select the architectural features (e.g., finish materials) and major building systems (e.g., laboratory equipment). Specific choices may be made by reviewing manufacturer documentation and by considering consultant recommendations. At this point in the process, the A/E team's knowledge and experience in forensic science laboratory design is critical. Wrong choices have significant cost and schedule impacts. Users must provide feedback and decisions to determine features and systems selections.

⑦ <u>**Refine/Value Engineer:**</u> The goal of this step is to align the plans with the project budget. The A/E team engages its SMEs, such as a CM, if on the team, to find opportunities to optimize the project's budget against the scope and site. The A/E team continues to refine the details of the major systems and each concept's functional capabilities. During this step, the project team answers all queries from the A/E team.

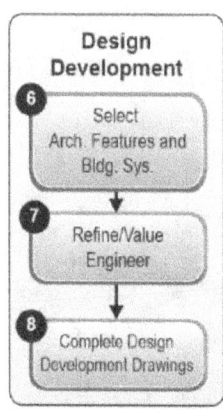

⑧ <u>**Complete Design Development Drawings:**</u> The A/E team incorporates all design decisions into design documents to complete the Design Development Drawings (deliverable). Project detailing is brought to the level needed to generate an accurate construction estimate. The A/E team presents the Design Development documents and the Project Cost Estimate (deliverable) for the users' review and sign-off. The deliverable includes a set of drawings that are approximately 2/3 complete as well as an estimate, which provides a close approximation of the final cost. The users engage project teams to validate the design documents and provide feedback and accept designs as appropriate.

Figure 3-5: Design development stage of the Design Phase.

Design Development Stage Best Practices

- Incorporate sustainable standards. Many owners have adopted the green building rating system for sustainable certification developed by the United States Green Building Council's (USGBC's) LEED, one of a few sustainable rating systems used throughout the country. LEED provides building users and operators with a measuring tool for identifying and implementing practical green building design, construction, operations, and maintenance solutions throughout a building's life cycle. Using LEED ensures that sustainable approaches are considered in new building and renovation projects. Designing a building to LEED standards increases the overall cost to build the facility approximately 2 or more percent. Building a facility to LEED standards may result in total life cycle savings. Life cycle savings have been estimated at 20 percent of total construction cost in today's dollars.

- Apply life cycle cost analyses when considering using new and emerging technologies to help justify new laboratory requirements (i.e., operational savings using more efficient technologies, such as occupancy sensors for fume hoods and lights).
- Take advantage of the design development stage as the best opportunity to minimize change orders.

3.2.4 CONSTRUCTION DOCUMENTS

Construction documentation takes the design development solution and translates it into a format that contractors can use to bid and ultimately construct the project. The construction documentation, shown in figure 3-6, includes both specifications and drawings. The drawings define the physical placement of materials, and the specifications define material quality.

⑨ Complete Drawings and Specifications: The A/E team prepares Construction Drawings (deliverable). Design development drawings are further refined and detailed so drawings can be used for bidding and construction. Site plans are finalized, and project detail drawings are prepared. The final cost estimate is also provided with the drawings. Users approve final documents or request corrections as necessary to match the project's intent and to fit within resources. The A/E team uses the results from the construction drawing completion to finalize Construction Specifications (deliverable) for all materials. The A/E team then prepares documents for

Figure 3-6: Construction documents stage of the Design Phase.

bidding construction. Users approve final specifications or request changes to each as necessary to match the project's intent and to fit within resources.

Construction Documents Stage Best Practices

- Respond rapidly to information requests/design clarifications during the construction documentation process. These requests are on the critical path for deliverable preparation.
- Seriously consider incorporating computerized capability to manage the completed facility. Consult your internal facility management team and consultant SMEs for options and advantages.

FAQ

Q: *What level of involvement should the laboratory director expect during the construction documents stage?*
A: Decision making during construction documentation is highly detailed and occurs at a fast pace. The level of engagement depends on the number of decisions that need to be made, which varies by project and project team. However, the overall success of designing a project to meet the user's needs is often proportioned to the level of engagement of the user team in the design process. The timeframe within which the user needs to respond remains consistent. Information requests and decisions require responses within 1 to 2 days.

3.3 DESIGN PROCESSES AND TOOLS

This section presents tools, technologies, and processes that the user may require during the Design Phase. Awareness of these elements can increase the impact of the user's participation. Also, please see appendix A: Standards, Policies, and Regulations for additional design resources.

3.3.1 BUILDING INFORMATION MODELING (BIM)

BIM, a relatively new technology within the building industry, uses an intelligent, model-based process to collaboratively construct a digital representation of physical, functional, and general building characteristics. This virtual facility model is shared and developed between the design team, constructors, the owner/operator, and any other key stakeholders in a collaborative and integrated environment that can improve productivity and project visualization, increase coordination among architects, engineers, contractors, and other stakeholders, and increase the speed and accuracy of project delivery. As a "shared" resource containing a myriad of facility information, BIM can be used throughout the facility life cycle, from inception to operation. It can also be a highly effective, useful tool for the A/E team to convey design ideas to the user group because three-dimensional images are easier for the layperson to read and understand than traditional two-dimensional plans and elevations.

Although BIM offers many advantages within the facility life cycle, using BIM and an integrated design process requires both a more sophisticated understanding of the facility delivery process and an A/E team experienced in using BIM to deliver projects. Furthermore, the use of BIM requires changes to the owner's decision making since, typically, using BIM requires decisions to be made much earlier in the design process. Owners considering the use of BIM to facilitate project delivery should carefully select project team members, designers, and contractors/subcontractors with a record of successful project delivery using BIM.

3.3.2 TECHNICAL LABORATORIES

Appendix B: Laboratory Design Considerations outlines types of technical laboratories and expands to include current forensic science laboratory departments (such as toxicology, biological science/DNA, firearms analysis, trace evidence, and chemistry). This appendix is a guide for technical requirements during the design process.

3.3.3 DESIGN PHASE CHECKLIST

The following checklist in table 3-2 identifies the activities that must be completed as part of performing the Design Phase. Checklists for all phases are provided in appendix F: Key Activity Checklists.

Table 3-2: Design Phase Checklist
(Note: bolded items are Phase deliverables)

	Process Step	Activity
☐	1	External A/E Consultant Identified
☐	1	Funding Approval Secured for A/E Consultant
☐	1	A/E Consultant SOW/RFP Defined
☐	1	A/E Consultant Contracted
☐	2	Design Program Reviewed with A/E
☐	2	Design Program Revised/Updated
☐	3	**Design Charrette** Executed
☐	4	Preliminary Concept Developed
☐	5	**Schematic Design** Selected

Process Step	Activity
☐ 6	Major Systems Selected
☐ 7	Design Value Engineered
☐ 8	**Design Development Drawings** Completed
☐ 8	**Project Cost Estimate** Completed
☐ 9	**Construction Drawing** Completed
☐ 9	**Construction Specification** Completed

Design

4. CONSTRUCTION

This phase transforms the laboratory plan and design into a constructed building. During the Construction Phase, a team of individuals represents the laboratory's interests to ensure that all the laboratory concepts are fulfilled. The laboratory needs to rely on the project team members' collective expertise with respect to interpretation of the contract documents, resolution of technical construction issues, and unforeseen construction conditions. Much of the team's time is spent in this phase inspecting and checking the work to ensure the laboratory is getting what it is paying for and responding to information requests. The nature of the project and available resources frames the key choices that are in the best interest of the laboratory. These choices include what type of construction delivery method to use to complete the project (e.g., design-build or public-private-partnership) as well as which contract type to use (e.g., fixed price or cost plus fee).

4.1 CONSTRUCTION PHASE ROLES AND RESPONSIBILITIES

During the Construction Phase, the majority of the time and effort occurs with the physical construction of the facility. As shown in table 4-1, the user group's primary role is to observe the work to ensure it meets expectations. The technical manager is the main point of contact for the general contractor. The project team must direct questions and comments through the

> "*Communication is essential. Weekly meetings during the construction process are a must. Users and A/E's must be responsive to contractor RFIs, and answers must be timely.*"
> — Laboratory User

technical manager. The A/E team also assigns a single point of contact to respond to general contractor questions, change orders, and design intent. The A/E team lead responds to the general contractor through the technical manager.

After acceptance of the final constructed product, the team engages in the steps of ensuring proper function and use of the laboratory.

Table 4-1: Project Team Activities during Construction Phase

Group	Role	Construction Phase Responsibilities
User (internal to laboratory)	Laboratory Director	Secures funding for construction. Selects the contractor team to perform the work. Accepts the final constructed product.
	Technical Manager	Manages day-to-day activities with a focus on continuous communication with the A/E team. Participates in regular in-progress construction inspections. Oversees commissioning and effective warranty evaluation. Ensures staff training. May be supported by an owner's representative during the laboratory construction.
	Project Committees	Maintains awareness of project and provide rapid responses to information requests. Participates in construction inspections. Receives operations training.
	Contracting Agent	Provides guidance on configuration of acquisition. Performs acquisition process for user group and execute contract. Tracks acceptance of work and payments.
Property Owner	Facilities Project Manager	Maintains constant contact during construction process to ensure effective installation into building. Conducts training of staff during commissioning for ongoing maintenance.
Design Team	Forensic Science Consultant	Provides recommendations for managing construction issues as they arise (e.g., assists with selecting general contractor-proposed equipment, request for information [RFI] responses, and final punch list inspection).
	A/E Team	Provides post-construction award services for facility. Prepares submittals and provides construction oversight. May provide commissioning agent responsible for leading the commissioning step of the process. Develops construction documents. Provides responses to contractors' requests for clarification, change order requests, and field clarifications.
Construction Team	General Contractor	Performs the construction. Depending on configuration of construction team, may include a CM. Participates in warranty evaluation.
	Subcontractor	Includes vendors with specialty construction capabilities. Reports directly to the general contractor. Does not engage user group or A/E team.

4.2   THE CONSTRUCTION PROCESS

From the user group's standpoint, there are three primary stages during the Construction Phase (figure 4-1): initiation, construction, and post-occupancy activities. Each stage requires a different set of decisions and different levels of participation by the user group.

> "On large projects, hire a good construction manager who will act as the owner's agent to make sure you are getting what you paid for in your project. This manager can help keep your change order costs down and keep your contractor honest."
>
> —Laboratory User

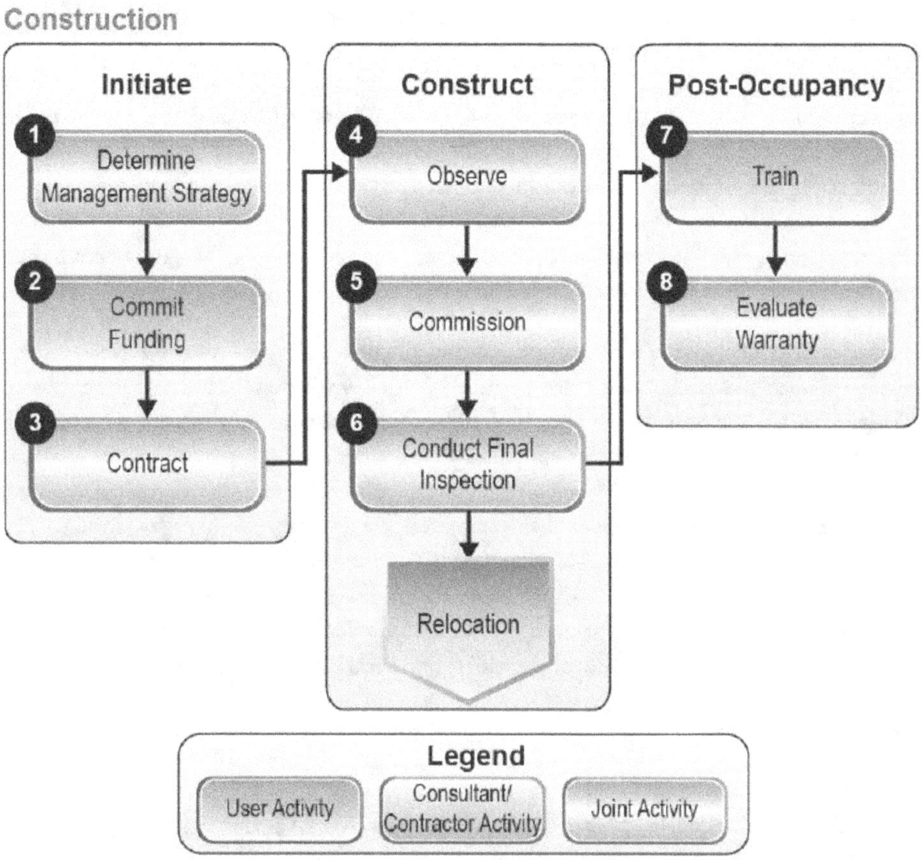

Construction
Initiate
1 Determine Management Strategy
2 Commit Funding
3 Contract
Construct
4 Observe
5 Commission
6 Conduct Final Inspection
Relocation
Post-Occupancy
7 Train
8 Evaluate Warranty
Legend
User Activity
Consultant/ Contractor Activity
Joint Activity

Figure 4-1: Forensic science laboratory project Construction Phase.

During each stage, construction team members have significant roles and responsibilities. Cooperation and coordination among team members are key to project success.

4.2.1 INITIATE

Key decisions made during the initiate stage, as shown in figure 4-2, require the direct attention of the user group.

Initiate
1 Determine Management Strategy
2 Commit Funding
3 Contract

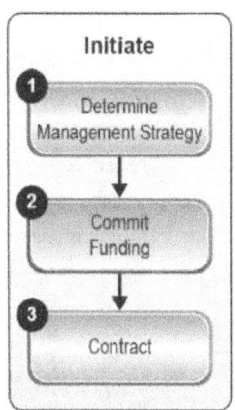

Figure 4-2: Initiate stage of the Construction Phase.

❶ **Determine Management Strategy:** This step requires the choice of construction team configuration and contract type. Developing solutions for these decisions can begin during the Planning Phase. The A/E team and contracting agent can provide advice on each of these choices. The laboratory director determines the construction team configuration. With the contracting officer, the laboratory director/technical manager also determines the contract type. (See section 4.3, Construction Processes and Tools, for more information.)

❷ **Commit Funding:** The technical manager needs to ensure the availability of funding required for construction. This is distinct from budgeting and programming funding for the project. The process of committing the funds

may require at least one month to complete. The contracting agent can assist with identifying processes and deadlines for committing funds to the project.

 ❸ Contract: This step includes the bidding process and Contract Award (deliverable). The contracting officer takes the lead on this step, but depending on team and contract configuration, the A/E team may lead contracting the construction of the forensic science laboratory. The bidding process is complex and requires the technical manager's participation. The process includes solicitation, a pre-bid conference, document issuance, RFI responses, addenda issuance, and bid opening. The user group must be available to the contracting officer for assisting with RFIs and addenda.

The award process includes the analysis of bids and alternates. All bids must be evaluated carefully. Often a project includes alternates, which are portions of the work that may or may not be completed. An additive alternate is additional work that the building owner may elect to do if the bid cost is within his or her budget. Multiple alternates in the bid package add additional complexity to bid evaluation. Once the successful bidder has been identified and the bid certified as bona fide, the construction contract between the user group and contractor can be completed. The technical manager should ensure that the appropriate signers of the contract are identified. Often, the contracting agent can sign the contract, but the laboratory director may also be required to sign. The laboratory director and the technical manager should assist in the final determination of a successful bidder (or make this determination solely if the project team does not include supporting entities) and should ensure that a contract is awarded.

 ## Initiate Stage Best Practices

- Map funding acquisition well in advance (one year) to ensure the project schedule. Depending on the user group's organization, the actions necessary to line up funds may have long lead times.

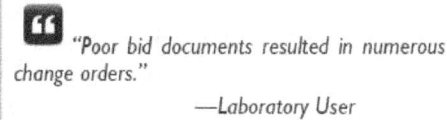
> *"Poor bid documents resulted in numerous change orders."*
> —*Laboratory User*

- Recognize that the purchase and installation of fixed or major equipment can be complicated and may result in coordination issues. The A/E team works with the contractor and his or her subcontractors to resolve these issues but may reach out to the technical manager and user group for clarification and direction.
- Address issues affecting existing equipment that will be relocated during laboratory construction. This includes moving the equipment, connecting the equipment, and potentially replacing equipment. The technical manager must be aware of any changes in equipment that may affect the building or utility requirements and communicate this to the A/E team for coordination with the contractor. Additionally, moving equipment could impact existing laboratory operations. The laboratory must make the decision to move only items that it can do without for a period of weeks (for small items) to months (for larger or more complex items that must be installed early in the Construction Phase).
- Work with the contracting officer to require bidders to provide performance bonds and past performance qualifications in preparing laboratory space to mitigate the risk that the low bidder cannot adequately perform the work.

FAQ

Q: *What is the difference between a construction manager and an owner's representative?*
A: Primarily, an owner's representative does not take responsibility for delivery of the built work, but ensures only that the project meets the user group's goals and expectations. From this standpoint, the owner's representative targets project management and assists with making the best decisions on behalf of the agency. Both roles report directly to the technical manager.

4.2.2 CONSTRUCT

The user group's primary goals during the construct stage, shown in figure 4-3, are to ensure the laboratory is built to meet design criteria and to accept the final built product.

④ Observe: The technical manager has two main tasks in this step: tracking project progress and assisting with construction deliverables as required. Tracking project progress includes ensuring that the project is on time and on budget. (See section 4.3, Construction Processes and Tools, for tools to assist with this task.) The nature of construction projects includes changes that challenge project resources. The construction team takes the lead on ensuring successful timing and budgeting. The laboratory director should maintain visibility on the main project metrics of schedule and budget and should be available to the construction team to provide decisions about changing circumstances.

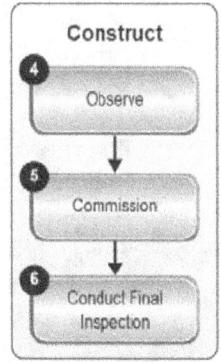

Figure 4-3: Construct stage of the Construction Phase.

Construction projects also require the user group to assist with several documents during the project life cycle, including submittals, change orders, and invoices. The user group works with the technical manager and A/E team in reviewing certain submittals and change orders associated with specific laboratory functions. A change order is a change to the contract that may require the user group's approval; for example, a change order can formalize the substitution of a piece of laboratory equipment. Invoices are typically submitted and processed monthly. Laboratory policies identify the user group's role in the invoicing process.

The construction team and the A/E team are responsible for processing these deliverables. There are instances in which the technical manager or the laboratory director may be required to provide direction on managing these deliverables (e.g., approval of a change order that significantly impacts schedule and/or budget). The user group must be available to the construction team to provide decisions on substantial deliverables that impact schedule or budget. The forensic science consultant participates in this stage by reviewing submittals and RFIs specifically related to the forensic design and by providing periodic observation of the construction progress related to the construction and installation of forensic design systems.

> *"Have a good understanding of the plans. Tour the site regularly. Be able to identify problems before they became huge cost issues. If you see something that does not look right, speak up."*
>
> —Laboratory User

⑤ Commission: During this step, the facility is commissioned to ensure that all equipment is fully operational. Because of the sensitive functions that occur within the forensic science laboratory, it is critical that the commission agent confirm the proper functioning of installed building systems and equipment. The technical manager also arranges for service of the equipment and systems at this time. Commissioning is the process of inspecting, testing, starting up, and adjusting building systems and then verifying and documenting that they are operating as proposed and meet the intended design criteria. This process expands on the customary testing, adjusting, and balancing typically performed on mechanical systems through a greatly broadened scope over a longer period of time. The process also ensures that all systems are installed properly and perform according to design, that they can be operated cost effectively, that they meet the user groups' needs, and that their operation is adequately documented and well understood by operators. The user group must ensure that the commissioning team understands and can confirm that the facility functions to meet user group requirements.

⑥ Conduct Final Inspection: The general contractor, with the A/E team, prepares a Punch List (deliverable) of outstanding items requiring construction completion. The technical manager conducts the final inspection with the general contractor's representatives and appropriate members from the user group (e.g., project Committee members). The inspection focuses on validating that all Punch List items have been resolved. The final invoice will then be processed by the contracting agent. Retainage (2 percent to 5 percent depending on contract terms) will be reserved by the user group for a specified time after final invoice payment (frequently one year). The user group must participate in the final inspection to ensure that work has been completed to the laboratory's satisfaction.

 If specified by the contract, at or prior to the final inspection, the user group should receive a set of project documents. Final inspection project documentation includes As-Built Drawings and Facility Equipment Owner's Manuals (deliverables). The A/E team works with the general contractor to provide as-built drawings and to deliver them to the user group for future reference. The construction team also provides the user group with O&M manuals for the major systems installed during the project, such as heating ventilation, and air conditioning (HVAC) systems and diagnostic equipment, for use by the user group's O&M staff and mission personnel. The technical manager must ensure the receipt of project documentation. The forensic science consultant participates in the final inspection process to ensure that forensic design systems conform to the construction documents.

At the end of the Construction Phase, the building is accepted by the governing agency and laboratory director. The user group staff members commence their move to the facility during the Relocation Phase. However, the Construction Phase continues with the post-occupancy stage, which entails the final steps required to ensure proper functioning of the laboratory.

 Construct Stage Best Practices

- Pay special attention to submittals requesting equipment substitutions to avoid cascading errors aligning with the overall project design. Confirm project specifications and clearly define the general contractor's responsibility for all corrections and adjustments resulting from requested product substitutions.

- Conduct regular construction progress meetings, which provide a standard construction industry control tool. The technical manager is encouraged to attend regularly. The meetings offer a chance for construction team members to address issues of concern and upcoming key tasks.

- Use a task tracking tool, frequently arranged as a database, to track issues that require resolution. A dashboard feature also can be implemented that provides a summary of task status to all team members. These tasks are then typically closely monitored in the weekly project meetings.

- Be aware that utility installation is often mishandled during construction. Additional attention and coordination applied to utility installation optimizes the construction effort. The organization of utility-distributed systems is critical for adaptability and ease of maintenance in laboratory facilities.

- During the construction process, be aware that a number of evolving issues may lead to revisions in the laboratory build-out. For example, it is not unusual for a large percentage of the laboratory users to change. The laboratory targeted program may change, or the construction process may reveal unresolved design decisions. Each of these situations may require design revisions. A significant amount of time may be required of the technical manager to coordinate these revisions during construction.

- Know that change orders will be inevitable.

- Do not move into the new facility until receiving the certificate of occupancy, performing a final inspection, completing the punch-list, and formally accepting the building from the general contractor. It is legal to move into a building with only a certification of substantial completion, which can be issued prior to a certificate of occupancy, but this is strongly discouraged. This introduces the risk of the contractor claiming any deficiencies in the building are the result of damage by laboratory personnel.

FAQ

Q: *What are the greatest risks specific to forensic science laboratory construction?*
A: Schedule and budget are the biggest risks to forensic science laboratory construction. Change orders are the biggest threat to keeping the project under budget. The laboratory director and technical manager can manage these with sound design work and tracking tools. Inexperienced contractors and unqualified low-bidders introduce additional risk. The A/E team and contracting agent can provide assistance with screening and selecting qualified contractors.

4.2.3 POST-OCCUPANCY

Post-occupancy is the 12-month period of time following the acceptance of the laboratory by the user group. During this time, as shown in figure 4-4, the user group ensures that the building can be operated as expected and that all systems under warranty are operating as required.

7 **Train:** The technical manager ensures that the O&M staff is properly trained to manage the facility and that laboratory staff is capable of operating the pertinent equipment. The technical manager obtains training for the facilities engineer and custodial staff on new equipment so they understand how the building systems work. The technical manager ensures that the O&M staff is aware of systems requiring user group support and that they possess the skills to provide this support. The technical manager must also ensure that laboratory staff members possess the ability to use laboratory assets.

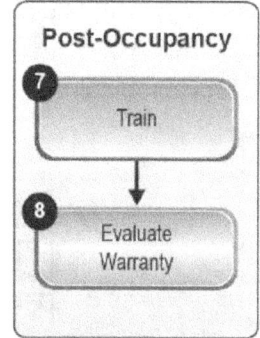

Figure 4-4: Post-occupancy stage of the Construction Phase.

8 **Evaluate Warranty:** Major facility components will possess a 12-month warranty. This step includes the inspection and correction of any systems functioning below specification. All parties involved in the Construction Phase participate in the successful performance of this step: the A/E team, the forensic science consultant, the commissioning agent, the general contractor, the facility owner, and the user group. The technical manager must maintain continual awareness of major components operating out of specification and must document these incidents for the A/E team. The user group should also participate in warranty inspections and ensure that all issues are identified and reported within the 12-month warranty window. The technical manager must follow-up to ensure all issues are resolved.

Post-Occupancy Stage Best Practices

- Assign a staff member to track all warranty-related performance issues for all major laboratory components for resolving during the warranty stage. Leverage BIM software for tracking if available.
- Verify that the project specifications include requirements for the general contractor and qualified equipment representatives to include training for laboratory users and O&M staff. Project specification should indicate items such as the following: a list of building system or equipment training that is to be provided for, the duration and number of training sessions, and a video recording of each training session.

4.3 CONSTRUCTION PROCESSES AND TOOLS

This section presents tools and processes the user group may encounter during the Construction Phase. These elements can increase user group impact. Also, appendix A offers additional construction resources.

4.3.1 CONTRACT TYPE AND PROCUREMENT OPTIONS

The type of contract selected for executing the construction work determines how the technical manager and the laboratory director will be required to participate in the project. This is primarily driven by how the contract will define the team that completes the project. There are several basic contract type options. Each provides a different way to manage the risk of successful project delivery. The lower the risk, the more precise and objective the terms of the bidding evaluation can be. Appendix C presents a review of some of these procurement options.

4.3.2 CONSTRUCTION PHASE CHECKLIST

The following checklist, in table 4-2, identifies the activities that must be completed as part of performing the Construction Phase. Checklists for all phases are in appendix F: Key Activity Checklists.

Table 4-2: Construction Phase Checklist
(Note: bolded items are Phase deliverables)

	Process Step	Activity
☐	1	Construction Team Configured
☐	1	Contracting Method Selected
☐	2	Construction Funding Committed
☐	3	Construction Contract Developed
☐	3	**Construction Contract** Awarded
☐	4	Laboratory-Related Submittals Reviewed
☐	4	Construction Process Tracking Planned
☐	4	Material and Systems Submittals Reviewed
☐	5	Facility Commissioning Validated
☐	6	Final Inspection Conducted
☐	6	**Punch List** Completed
☐	6	**As-Built Drawing**, Major System Manuals Received
☐	7	Facility Operations Staff Trained
☐	8	Warranties Evaluated and Documented

As science-focused organizations restructure in an era of consolidation and outsourcing, the need to move research and testing equipment, samples, and even complete laboratories to new locations increases. Government agencies and other organizations seek to optimize use of capital investments or shift analysis to locations with lower costs of doing business. Forensic science laboratories, regardless of size or specific field, can minimize the financial, scientific, and regulatory compliance risks associated with relocation if the appropriate steps are taken. Certainly, the overarching goals should be to streamline the transition process, to ensure the safety of instruments, evidence, and occupants, to maintain a laboratory's compliance standing, and to provide the laboratory director with real-time visibility throughout the process.

Relocation requires deliberate, careful, and documented steps that must be addressed in a manner that ensures maximum uptime at both ends of the process. Instrumentation needs to be inventoried, and decisions need to be made about what to move (versus what to retire) and how to best deploy instrumentation at another facility. Protocols must be followed for decommissioning and re-commissioning. Significant care must be taken in preparing and transporting scientific instruments and valuable samples.

In addition, laboratory work must continue even as the shift is made from the old facility to the new. Laboratory directors must develop a strategic plan for the move. They must phase and schedule relocation activities so that routine daily events occur as smoothly and efficiently as possible. Every laboratory move has its own unique workflow and uptime sequencing requirements. Planners should therefore consider all relevant tasks and activities, the order in which these need to be accomplished, and the responsible party for each task. The relocation plan can then be divided into pre-move, move, and post-move stages.

5.1   RELOCATION PHASE ROLES AND RESPONSIBILITIES

The move process requires a high level of engagement by all laboratory staff members as outlined in table 5-1. External resources are limited to a vendor that provides moving services.

Table 5-1: Project Team Activities during Relocation Phase

Group	Role	Relocation Phase Responsibilities
User (internal to laboratory)	Laboratory Director	Maintains visibility of relocation process. Approves acquisition of move services.
	Technical Manager	Manages day-to-day actions and decisions for completing the relocation. Can be re-designated move captain or can assign a move captain responsible for relocation activities (as further detailed in this chapter).
	Project Committees	Provides input on detailed decision making for move process.
	Contracting Agent	Contracts the services of a moving company, as needed by the user group.
Property Owner	Facilities Project Manager	Provides coordination to support move.
Design Team	*Forensic Science Consultant*	*Does not participate.*
	A/E Team	*Does not participate.*
Construction Team	*General Contractor*	*Does not participate.*
	Subcontractor	*Does not participate.*

5.2 THE RELOCATION PROCESS

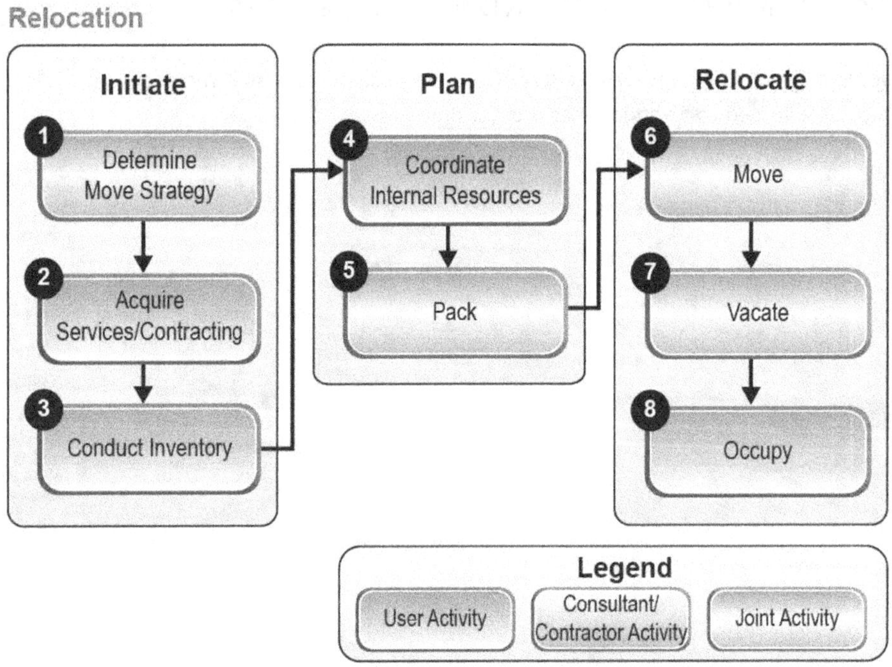

Figure 5-1: Forensic science laboratory project Relocation Phase.

5.2.1 INITIATE

From the beginning of the Relocation Phase figure 5-1, the user group must be directly involved in the process. A move captain is designated to oversee the activities. Figure 5-2 outlines the initiate stage.

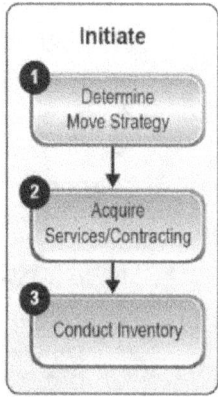

Figure 5-2: Initiate stage of the Relocation Phase.

❶ **Determine Move Strategy:** Moving is a complex series of tasks that require constant communication and cooperation. The user group should plan a series of meetings and activities preceding the relocation so that all details are planned out. These meetings incorporate facility inventories developed during the Design Phase. The resulting deliverable is the Integrated Master Schedule. This tool defines each distinct project activity and the critical path tasks that, if delayed, delay the completion of the project.

Another deliverable in this step can be a Task Tracker Tool. This tool, frequently arranged as a database, tracks issues that require resolution. A dashboard feature also can be implemented that provides all team members with a summary of task status. These tasks are closely monitored in project meetings.

The user group should attend all construction meetings and should keep the project team well informed of schedule changes. The user group should also generate a realistic laboratory move schedule, assess the site, and evaluate the documentation compliance needs relevant to post-installation qualification and validation.

A key consideration in a move strategy is the budget for the move. Move planners must consider lining up funding years in advance of the move. Budget should include a contingency for scope changes during project lifetime.

❷ **Acquire Services/Contracting:** This step includes the bidding process and contract awards for the services required. The user group must identify the types of services it requires. These services may include IT and furniture vendors, move managers, and move contractors. The contracting officer takes the lead on

> 66 *"We have always used some form of contracted move management services for our relocations."*
> — *Laboratory User*

these steps by determining the type of contracts to be used and the bidding processes to be followed.

During the bidding process, qualified bidders are invited to prepare bids for the move of the forensic science laboratory. The bidding process includes multiple steps requiring the technical manager's participation. These steps include solicitation, document issuance, a bid conference, question responses, and bid award. The user group should be available to the contracting officer for assisting with questions and addenda.

The award process includes the analysis of bids and alternates. All bids must be evaluated carefully. Often a project includes alternates. Alternates are portions of the work that may or may not be purchased or executed. Multiple alternates in the bid package add additional complexity to bid evaluation. Once the successful bidder has been identified, the contract between the user group and vendor can be executed. The contracting officer can sign the contract, and, depending on the situation, the contract may also need to be signed by other laboratory staff members.

The user group should assist in final determination of successful bidder (or make this determination solely if the project team does not include supporting entities). The user groups should also ensure that the contract is executed.

③ Conduct Inventory: The user group should begin planning the move by taking inventory of what is already on hand and what needs to be disposed of. This Inventory (deliverable) may build on those previously developed in the Design Phase. The group should analyze

> "*Planning, Planning, Planning – that was the key to our [relocation] success.*
> — *Laboratory User*"

hazardous materials, understand local regulatory compliance procedures, and review codes and laws relevant to the laboratory materials with which they work. The user group can contact equipment manufacturers for recommendation on equipment management during the move.

To maintain the warranty on certain instruments, utilize service technicians supplied by the instrumentation manufacturer. The user group should check existing service contracts for equipment maintenance to see if they include moving support.

The user group should contact the local authority for proper handling of hazardous materials waste, chemicals, etc. The group should establish disposal procedures and distribute them to the laboratory staff members.

Initiate Stage Best Practices

- Work with the contracting officer to require bidders to provide past performance qualifications in preparing laboratory space to mitigate the risk that the low bidder cannot adequately perform the work.
- Hire or designate a move captain who must have an overall understanding of the laboratory culture to grasp the scope of the project. The move captain must be able to communicate with the scientific community to recognize and translate their needs and must comprehend and be able to communicate the equipment and chemical requirements to contractors, vendors, and trades people. The move captain coordinates the move, oversees the subcontractor's schedules, attends weekly construction meetings, and reconciles laboratory's equipment needs. The move captain also possesses a thorough knowledge of the construction process.
- Keep all stakeholders involved during the initiate stage.
- Ensure that the user group uses project tracking and scheduling tools created with the project team to coordinate subcontractor's contracts, meetings, and timelines.
- To streamline the move, consider using a radio frequency identification (RFID) tracking system to inventory all devices and items that will be relocated. The RFID system can identify the item, retain critical notes for devices, note the room where item is currently located and the room where it will be located, and track all personnel handling the item and vehicles transporting the item.

5.2.2 PLAN

④ Coordinate Internal Resources: As shown in figure 5-3, the move captain ensures that specific service requirements for the move are met. The service requirements include, but are not limited to, IT; furniture, fixtures, and equipment (FF&E); and security.

Information Technology: Prior to the move, the user group should coordinate with IT support to determine whether the internal IT team or the contracted movers are responsible for moving IT equipment. Typically, the moving contractor does not move IT equipment unless it is unplugged/disassembled and ready to be moved, and he or she does not usually reconnect the equipment at the destination. Items to be coordinated include the network, telecommunications/local area network closet, cable management, utilities, and audio/visual support and connectivity.

The user group should contact IT support to disconnect the equipment prior to the scheduled move date and to reconnect it at the destination location. The group should also notify the movers of the requirement to move the IT equipment at the time the move is scheduled so that the movers bring the appropriate special equipment necessary to move IT equipment safely.

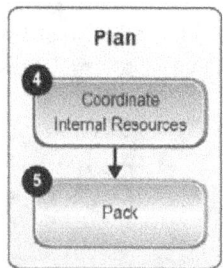

Figure 5-3: Plan stage of the Relocation Phase.

Furniture, Fixtures, and Equipment: FF&E includes vendors and services. Prior to the move, the user group should coordinate with the facilities project manager to determine the delivery and installation schedule. The schedule should synchronize with the general contractor's overall project schedule. After the furniture is installed, IT support makes the network connections. The vendor prepares a Punch List of outstanding items. The technical manager completes the final inspection with the facilities project manager and the vendor's representatives.

The facilities project manager coordinates with the vendor and building owner to manage the furniture delivery and installation. In most instances, the facilities project manager coordinates the arrival schedules, the number of trucks to be parked simultaneously at the destination site, and the use of any loading docks and freight elevators where applicable. Moves should be scheduled to comply with any building restrictions that may be in effect.

The user group should participate in the final inspection to ensure specifications are met and installation has been completed to the laboratory's satisfaction.

Security: Depending on the specific security requirements of individual sites, on the day prior to the move, the movers may be required to be escorted through the facilities or to submit personal information for clearance to obtain visitor passes. Move captains should be aware of and follow the access procedures for the particular site where access is required.

⑤ Pack: Laboratory directors should refer to the laboratory relocation guidelines in appendix A to safeguard and pack chemicals and equipment. Items are tagged with the destination location prior to the move, and laboratory directors/move captains advise the movers. The move captain should estimate the number of moving boxes needed and arranges for boxes and plastic moving crates to be delivered to the origination site approximately one week prior to the move.

The user group ensures that employees are aware of proper safety procedures for packing chemicals to be moved. The group provides and distributes laboratory relocation guidelines to employees and movers for protection and confirms the construction schedule.

Employees are generally responsible for packing up the contents of their offices/benches. The move captain procures packing boxes and delivers them to the current location. It is imperative that crates are properly identified to ensure safety of handling.

Plan Stage Best Practices

- Develop a "from-to" list, which contains information on each laboratory employee, to submit to the laboratory director or facilities manager.
- Use the relocation as an opportunity to write new internal business policies.

- Because movement of evidence is a key driver of shaping the move plan, establish chain of custody plan and escort requirements and identify physical locations for evidence throughout the move process. Prior to the move, notify law enforcement agencies, the district attorney's office, and other legal representatives of the plan to move evidence.
- Establish a building support and custodial team for the operation of the new facility.
- Prepare and implement a process for certifying, accrediting, and calibrating equipment once it has been moved to the new location.
- Plan to move all equipment over one weekend to minimize disruption to ongoing laboratory functions.
- Use the opportunity of the move to donate, recycle, or discard old or obsolete items in the laboratory. Provide a specific number of boxes for employees to use for their personal items, and encourage them to dispose of any items that do not fit in those boxes. Provide recycling and/or disposal for universal waste, including batteries, computer monitors, consumer electronic devices, mercury-containing devices, and fluorescent lamps.
- Arrange for law enforcement security during the movement of the laboratory's firearms.

FAQ

Q: *How far in advance should I contact a moving company to schedule employee relocation?*

A: Schedule the move as far in advance as possible, especially during the busy moving season from May 15 to September 15. Six weeks from the move date is not too early. Keep a close account of the construction schedule as moves may need to be rescheduled because of unforeseen construction and inspection delays.

5.2.3 RELOCATE

6 **Move:** The relocate stage is shown in figure 5-4. The user group confirms that IT systems, FF&E, and telecommunications tools are installed on the planned date. The move captain ensures the physical move of boxes, equipment, and other laboratory material occurs as scheduled.

7 **Vacate:** Refer to appendix G for checklists on properly ensuring safe and compliant transitions when vacating a laboratory facility. The user group should be familiar with the policies, training, and decommissioning standards required to vacate a laboratory. The group oversees the appropriate handling of waste materials and ensures that arrangements for this work are made well in advance to allow contractors sufficient time to meet the schedule.

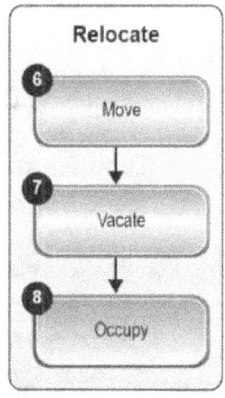

Figure 5-4: Relocate stage of the Relocation Phase.

8 **Occupy:** At this point, the facility is ready for move-in. The move contractor begins post-occupancy testing and calibration/validation of instruments on each system. O&M manuals are provided by the contractor.

The user group oversees the move, reinstallation of the equipment, and documentation of system validation as needed. The group schedules pick-up of moving crates and boxes and establishes maintenance procedures.

Relocate Stage Best Practices

- Identify issues that need to be addressed. Determine whether these issues will be addressed by an internal staff member or by an external consultant, because post-occupancy evaluations and feedback are essential for the continuous improvement of building laboratories. Prepare questionnaires, interviews, and focus groups. Determine a benchmarking process and decide how to share the information with the next project team or other laboratory directors.

- Ensure that construction is complete prior to commissioning. Commissioning should be complete before the move.
- When packing, take the opportunity to purge unused or unneeded files and equipment.
- Prepare document for laboratory staff with human capital factor best practices (e.g., best way to lift a box).
- Work with the A/E team to conduct post-occupancy evaluation because this feedback is critical to continued improvement in providing facility design for other forensic laboratory users.

FAQ

Q: *How do I engage the public with the new facility?*
A: Consider holding a building open-house for families, press, and the public. Highlight innovations and improvements in the new facility.

5.3   RELOCATION PROCESSES AND TOOLS

In addition to the following items, appendix A: Standards, Policies, and Regulations offers other resources.

5.3.1   VACATING A LABORATORY

Again, the goal of the Relocation Phase is to ensure safe and compliant transitions in laboratory occupancy. The four major areas to address when vacating a laboratory are general safety, biological safety, radiation safety, and chemical safety.

Laboratory directors should carefully review the handling and disposal of waste materials. Once hazardous materials are properly disposed of, they must ensure that all chemicals to be moved are properly labeled and that the containers are safe to handle. Broken or disintegrating containers of chemicals should be disposed of. Appendix G presents detailed requirements.

5.3.2   DECOMMISSIONING REQUIREMENTS

Decommissioning consists of all work required to achieve a facility free of chemical, biological, radiological, or other hazardous materials. This process prepares the area for reasonable, unrestricted demolition. Decommissioning includes an in-depth facility assessment by a qualified environmental engineer or industrial hygienist. The assessment identifies any environmental or other site hazards that could result in the release of hazardous substances during demolition or that could pose a hazard to workers. A checklist for managing these requirements is included in appendix G.

5.3.3   TRANSITION PROCESS

Relocation planning requires specific actions resulting in a detailed, realistic, working transition plan and budget that ensure that the move remains on schedule and budget with full operational capability in the new facility on day one. The transition plan includes the tasks, time frames, costs, and responsibilities for subordinate activities required to ensure a successful transition.

> "*Plan for more time. The move itself was not bad, but getting settled into the new site takes much longer than you anticipate.*"
> — Laboratory Director

Planners should understand the "baseline" to accurately estimate performance risk and to develop an efficient plan. To understand what the move entails requires capturing accurate data on the people, mission-critical and stored assets, personal property, equipment, systems, and data that will be impacted by the move. To define the full scope and to establish the baseline, the inventory should include, minimally, IT systems and subsystems, active hard-copy and archived records, paper and film storage, confidential material, and mission-critical equipment and assets.

1. **Requirements Development:** Using the baseline information that has been collected, a comprehensive analysis of transition requirements must be developed in coordination with the various stakeholder groups to ensure no degradation in mission. Preparing this analysis will require an understanding of mission-critical operations.

2. **Equipment Planning:** This process begins with meeting with stakeholders assigned to the project and other identified stakeholders to discuss requirements that support the final objective: outfitting the new facility with the equipment needed to support the organization's mission and operational requirements.

3. **Continuity of Operations and Concept of Operations:** The organization cannot stop functioning because of the relocation; therefore, considerations must be integrated into transition planning to ensure no disruption or degradation to mission accomplishment during and after the transition occurs. Prior to the move, notify courts and law enforcement agencies to avoid conflicts with potential trial needs.

4. **O&M Planning:** The planning process ensures that the new facility is ready for operation on day one and integrates considerations of the new facility's O&M once the relocation has been completed.

5.3.4 RELOCATION PHASE CHECKLIST

The following checklist, in table 5-2, identifies the activities that must be completed as part of performing the Relocation Phase. Checklists for all phases are in appendix F: Key Activity Checklists.

Table 5-2: Relocation Phase Checklist
(Note: bolded items are Phase deliverables)

	Process Step	Activity
☐	1	Move Captain Designated
☐	1	**Integrated Master Schedule** Prepared
☐	1	**Task Tracker Tool** Assembled
☐	2	Contracted Move Service Requirements Identified
☐	2	Contracted Services Acquired
☐	3	**Equipment Inventory** Prepared
☐	4	Move Responsibilities Coordinated
☐	5	All Materials Packed
☐	6	New Site Material Installations Coordinated
☐	7	Old Laboratory Vacated to Manage All Applicable Compliance Issues
☐	8	Equipment in New Location Calibrated/Validated

Process Step	Activity
8	Staff Moved into New Laboratory

Additive Alternate – Additional work that increases the cost of a construction bid.

Architect-Engineer Services (as defined in 40 USC 1102 and the Federal Acquisition Regulation [FAR]) –

A. Professional services of an architectural or engineering nature, as defined by state law, which are required to be performed or approved by a person licensed, registered, or certified to provide such services;

B. Professional services of an architectural or engineering nature performed by contract that are associated with research, planning, development, design, construction, alteration, or repair of real property; and

C. Such other professional services of an architectural or engineering nature, or incidental services, which members of the architectural and engineering professions (and individuals in their employ) may logically or justifiably perform, including studies, investigations, surveying and mapping, tests, evaluations, consultations, comprehensive planning, program management, conceptual designs, plans and specifications, value engineering, Construction Phase services, soils engineering, drawing reviews, preparation of operating and maintenance manuals, and other related services.

As-Built Drawings – Construction drawings revised to show specific detailed changes made during the construction process, based on record drawings (marked-up prints, drawings, and other data) furnished by the Contractor to the Government. The drawings clearly identify that they are the "as-built" drawings.

Basic Services – The services performed by an architect-engineer during the project: schematic design, design development, construction documents, bidding or negotiation, and contract administration.

Benchmarking – Process of analyzing and comparing facility design projects with industry standards.

Bona Fide – A bid that is submitted in good faith and compliant with bidding documents.

Certificate of Occupancy – A document issued by the governing jurisdiction permitting the owner to occupy the building for approved use and population.

Change Order – A revision to the contract for construction that is signed by the owner, contractor, and architect.

Commissioning – The process of inspecting, testing, starting up, and adjusting building systems and then verifying and documenting that they are operating as proposed and meet the intended design criteria.

Concept Drawings – Graphics showing alternatives used to define a project's scope during the programmatic stage of the project.

Construction Codes – Any set of standards set forth in regulations, ordinances, or statutory requirements of a Federal, state, or local governmental unit relating to building construction and occupancy, adopted and administered for the protection of the public health, safety, and welfare and of the environment.

Construction Contract (as defined by FAR, refer to the appropriate authority for local and state regulations) – A mutually binding legal relationship obligating the seller to furnish the supplies or services (including construction) and the buyer to pay for them. It includes all types of commitments that obligate the Government to an expenditure of appropriated funds and that, except as otherwise authorized by the FAR, are in writing. In addition to bilateral instruments, contracts include (but are not limited to) awards and notices of awards; job orders or task letters issued under basic ordering agreements; letter contracts; orders, such as purchase orders, under which the contract becomes effective by written acceptance or performance; and bilateral contract modifications.

Construction Documents Phase – A term used to describe many types of documents used to take a project from design to completed building. The third phase of the architect-engineer's basic services, where the architect-engineer prepares from the approved design development documents, for approval by the Government agency having jurisdiction, the working drawings and specifications and the necessary bidding information.

Construction Management – A professional service that applies effective management techniques to the planning, design, and construction of a project from inception to completion for the purpose of controlling time, cost, and quality, as defined by the Construction Management Association of America.

Construction Manager – A person, firm, or business organization with the expertise and resources that has the responsibilities under contract to act as the owner's agent and provide the owner with impartial technical advice.

Contract Documents – Those documents that compose a contract, such as in a construction contract, the Government contractor agreement (Standard Form 252, which includes general provisions and clauses; special contract requirements; other provisions in the uniform contract format; specifications, plans, and/or drawings; all addenda, modifications, and changes thereto, together with any other items stipulated as being specifically included).

Contracting Officer – An individual who has the authority to execute a contract on behalf of the Government agency having jurisdiction and to make changes, amend contracts, approve payments, terminate contracts, and close out contracts upon satisfactory completion. The sole authorized agent in dealing with the contractor.

Contracting Officer's Technical Representative – The project officer or other authorized representative who is designated by the contracting officer.

Contractor – The person, firm, or corporation with whom the Government has executed a contract and who is responsible for performing the work.

Critical Path – A method for scheduling a set of project activities. The uninterrupted sequence of path of project activities that represents the time duration from that point to project completion and often continually changes with time.

Decommissioning – A process that consists of all work required to achieve a facility free of chemical, biological, radiological, or other hazardous materials.

Design-Bid-Build – (as defined by FAR, refer to the appropriate authority for state and local regulations) – The traditional delivery method where design and construction are sequential and contracted for separately with two contracts and two contractors.

Design-Build – (as defined by FAR, refer to the appropriate authority for state and local regulations) – The delivery method that combines design and construction in a single contract with one contractor. Under this approach, the contractor and the A/E team are teamed to design and construct the facility under one contract.

Design Program – A deliverable during the Design Phase that translates requirements identified in the Needs Assessment into data that are used by architects and engineers to design the facility.

Fast-Track Construction – A scheduling process in which design and construction activities overlap. Design documents and equipment and trade subcontracts released incrementally or in phases.

Federal Acquisition Regulation – The basic policy governing Federal agency acquisitions that contains legal requirements, regulations, and policies that bear on contracting. The FAR is available at http://farsite.hill.af.mil/vffara.htm or http://www.acquisition.gov/comp/far/index.html.

Footprint – The physical "floor" space required for an object or set of objects.

General Services Administration (GSA) – U.S. Federal agency responsible for property acquisition and management. Federal organizations can expect to partner with GSA during many of the processes described in this handbook.

Gross Area – The total square footage/square meters in a building for all floors from the outside face of exterior walls, disregarding such architectural projections as cornices, buttresses, and roof overhangs. Gross area includes all research and administrative space; retail space; other areas, such as vending machine space and storage; and major vertical penetrations, such as shafts, elevators, stairs, or atrium space. This figure is used in defining construction costs for facilities.

Historic Properties – Properties listed on the National Register of Historic Places or determined by the Federal Preservation Officer with the cognizant State Historic Preservation Officer or Tribal Historic Preservation Officer to be eligible for listing on the National Register of Historic Places based on National Register Criteria.

Improvements (Renovations/Alterations) – Any betterment or change to an existing property to allow its continued or more efficient use within its designated purpose (renovation) or for use for a different purpose or function (alteration). Building improvements also include improvements to or upgrading of primary mechanical, electrical, or other building systems and site improvements not associated with construction projects.

Integrated Master Schedule (IMS) – A tool that defines each distinct project activity and the critical path tasks that, if delayed, delay the completion of the project. This tool can be utilized during the Relocation Phase.

Lease – Specific rights to real property that have been assigned to the Federal Government for a defined period of time. A Federal lease is both a conveyance and contract to possess and use real property for a pre-determined period of time.

Life Cycle Cost – The total cost of owning, operating, and maintaining a building over its useful life, including its fuel and energy costs, determined on the basis of a systematic evaluation and comparison of alternative building systems; except that in the case of leased buildings, the life cycle cost is calculated over the effective remaining term of the lease.

Model Building Codes – Regional building codes adopted as law by local jurisdictions.

Nationally Recognized Standard – Benchmark that encompasses any standard or modification thereof that

- has been adopted and promulgated by a nationally recognized standards producing organization, often called standards development organization (SDO), under procedures whereby those interested and affected by it have reached substantial agreement on its adoption, or
- was formulated through consultation by appropriate Federal agencies in a manner that afforded an opportunity for diverse views to be considered.

Needs Assessment – An essential planning tool, typically developed by independent professionals (e.g., forensic science consultants), that considers research, user, and facility needs; assesses the existing facility condition, limitations, and challenges; defines all space requirements; and refines preliminary opinions of construction cost.

Net Area/Net Space – Portions of the facility available to use for program operations and for supply storage, building maintenance/operation, and other necessary support functions. Net Area is measured from the inside of the permanent exterior wall to the near side of permanent walls separating the area from stairwells, elevators, mechanical rooms, permanent corridors, or other portions of the building not categorized as Net Space Area in the program of requirements document. In calculating net area, no deduction is made for columns and projections that are necessary to the building. However, deductions are made for large duct and elevator shafts passing through it.

Net Assignable Square Footage – The area of a floor or office suite that is suitable for occupancy, including secondary corridors and excluding common or shared space that cannot be reasonably assigned for program purposes, such as main egress corridors, hazardous waste marshaling areas on the loading dock, and other non-programmable space.

Pre-Project Planning – A process for developing sufficient strategic information used to address risk and determine required resources for successful construction projects.

Project Management Plan (PMP) – A deliverable that outlines the strategy for planning, designing, constructing, commissioning, and relocating into the facility. The PMP is a dynamic document that is continually updated and refined throughout the project life cycle.

Public Area – Any area of a building that is ordinarily open to members of the public, including lobbies, courtyards, auditoriums, meeting rooms, and other such areas

Punch List – A deliverable prepared at the end of construction defining outstanding items requiring construction completion. The general contractor does not receive final payment until the Punch List has been resolved

Record Drawings – The drawings submitted by a contractor or subcontractor at any tier to show the construction of a particular structure or work as actually completed under the contract.

Request for Information (RFI) – A question from the contractor regarding the drawings or specifications.

Request for Proposals (RFP) – A document an organization uses to solicit bids. It outlines the contract terms and bid formatting requirements.

Retainage – Money withheld from periodic payments made to the contractor for work installed, which is incrementally substantially completed but still in need of final completion. It is equivalent to the cost of making the work complete and provides the owner with financial leverage.

Schematic Design – A deliverable developed by the A/E team that translates the project scope into conceptual design drawings possessing all the functionality required of the project.

Scope of Work (sometimes, referred to as 'Scope') – The narrative description of a project including the physical size and characteristics, functions, and special features.

Stakeholders – Individuals and organizations that are involved in or may be affected by the undertaking.

Statement of Work (SOW) – A document in the acquisition process that describes the work to be performed or the services to be rendered, defines the respective responsibilities of the Government and the contractor, and provides an objective measure so that both the Government and the contractor know when the work is complete and payment is justified. Common elements of the SOW are background, project objectives, detailed technical requirements, deliverables, reporting, schedule, special considerations, and references.

Submittals – Submittals are used to validate the details of the general contractor's construction performance. Examples of submittals can include documentation of equipment to be installed (cut sheets) or a mockup of a building's façade materials.

Usable Square footage (also referred to as "office area") – The secured area (square footage) occupied exclusively by a tenant within a tenant's leased space. The useable area multiplied by the load factor for common area results in rentable area on which rent is charged. The useable area can be measured in many ways, but the most common measurement for office buildings is according to industry standards. The useable area does not include restrooms, elevator shafts, fire escapes, stairwells, electrical and mechanical rooms, janitorial rooms, elevator lobbies, or public corridors (for example, a corridor leading from the elevator lobby to the entrance of a tenant's office).

Value Engineering – Process aimed at reducing project costs while maintaining basic functions.

COMMONLY USED ACRONYMS

A/E	Architect/Engineer
AFIS	Automated Fingerprint Identification System
BIM	Building Information Modeling
CCTV	Closed-Circuit Television
CLIA	Clinical Laboratory Improvement Acts
CM	Construction Manager
CMc	Construction Manager at Risk
CODIS	Combined DNA Index System
DB	Design-Build
DBB	Design-Bid-Build
DEA	Drug Enforcement Administration
ECI	Early Contractor Involvement
EPA	U.S. Environmental Protection Agency
FAR	Federal Acquisition Regulation
FAQ	Frequently Asked Question
FF&E	Furniture, Fixtures, and Equipment
GSF	Gross Square Feet
GSA	General Services Administration
HVAC	Heating, Ventilation, and Air Conditioning
IBIS	Integrated Ballistic Identification System
IPD	Integrated Project Delivery
ISO/IEC	International Organization for Standardization/International Electrotechnical Commission
IT	Information Technology
LEED	Leadership in Energy and Environmental Design
LPTA	Low Price Technically Acceptable
NFPA	National Fire Protection Association
NIH	National Institutes of Health
NIJ	National Institute of Justice
NIOSH	National Institute for Occupational Safety & Health
NTP	Notice to Proceed
O&M	Operations and Maintenance
OLES	Law Enforcement Standards Office
PBS	Public Buildings Service
PMP	Project Management Plan
PPP	Public–Private Partnership
RFI	Request for Information
RFID	Radio Frequency Identification
RFP	Request for Proposals
SEM	Scanning Electron Microscope
SME	Subject-Matter Expert
SOW	Statement of Work
TWG	Technical Working Group
USGBC	U.S. Green Building Council
VA	Department of Veteran Affairs
WBDG	Whole Building Design Guide

Commonly Used Acronyms

This appendix includes a set of national and industry-wide standards for the design and operation of forensic science laboratories. State and local requirements must also be examined to ensure a compliant laboratory. In addition to these references, consider the following tools: forensics blogs/chat rooms, surveys from the American Society of Crime Laboratory Directors, industry associations, and the forensic professional community network.

PLANNING STANDARDS

Issuing Organization	Title	Description
Department of Veteran Affairs (VA)	TIL - Design Guides (PG-18-12)	The document is a guide and supplement to current technical manuals and other VA criteria in the planning of Research Laboratory facilities.
General Services Administration (GSA)	P-100 – Facilities Standards for the Public Buildings Service (PBS)	The document establishes design standards and criteria for new buildings, major and minor alterations, and work in historic structures for the PBS of the GSA. This document contains policy and technical criteria to be used in the programming, design, and documentation of GSA buildings.
International Association of Chiefs of Police	Police Facility Planning Guidelines	This desk reference assists police administrators to plan, design, and build facilities to meet present and future needs.
International Organization for Standardization	ISO/IEC 17025:2005	ISO 17025 is the primary standard used by testing and calibration laboratories. Laboratories use the standard to implement a quality system focused on improving the ability to consistently produce valid results. This standard is the basis for accreditation from an accreditation body.
National Institutes of Health (NIH)	NIH Design Policy and Guidelines	The document establishes policy, design standards, and technical criteria for use in programming, designing, and constructing new buildings, and major alterations for the NIH.
Scientific Equipment and Furniture Association (SEFA)	4th Edition SEFA Desk Reference	This reference offers recommended practices for the use and application of scientific furniture (e.g., fume hoods, laboratory casework)
U.S. Environmental Protection Agency (EPA)	A Design Guide for Energy-Efficient Research Laboratories	Laboratories for the 21st Century (Labs21) is designed to meet the needs of laboratory and high-performance facility designers, engineers, owners, and facility managers. Cosponsored by the EPA and U.S. Department of Energy, Labs21 offers professionals worldwide an extraordinary opportunity for information exchange and education through three interactive components: a Partnership Program, Training and Education, and an interactive Tool Kit.
US Green Building Council (USGBC)	Leadership in Energy and Environmental Design (LEED) Rating System	LEED is an internationally recognized green building certification system. Developed by the USGBC in March 2000, LEED provides building owners and operators with a framework for identifying and implementing practical and measurable green building design, construction, operations, and maintenance solutions. LEED promotes sustainable building and development practices through a suite of rating systems that recognize projects that implement strategies for better environmental and health performance.

Issuing Organization	Title	Description
Wiley	Architectural Graphic Standards	These standards present accepted architectural practices in a clear and accessible graphic form.
Wiley	Building Type Basics for Research Laboratories	This document presents design of research laboratories with the practical information to meet construction and renovation needs.
Wiley	Design and Planning of Research and Clinical Laboratory Facilities	This document provides a comprehensive overview of the fundamental issues related to laboratory design.
Wiley	Guidelines for Laboratory Design: Health and Safety Considerations	This document provides design information related to specific health and safety issues that need to be considered when building or renovating laboratories

DESIGN STANDARDS

Issuing Organization	Title	Description
Americans with Disabilities Act	2010 ADA Standard for Accessible Design	The 2010 Standards set minimum requirements—both scoping and technical—for newly designed and constructed or altered state and local government facilities, public accommodations, and commercial facilities to be readily accessible to and usable by individuals with disabilities.
Center for Medicare and Medicaid Services	Clinical Laboratory Improvement Act (CLIA)	The objective of the CLIA program is to ensure quality laboratory testing.
Department of Veterans Affairs	VA Research Laboratory Design Guide	The goal of the VA Design Guide is to ensure the quality of VA medical research laboratories while controlling construction and operating costs. This document is intended to be used as a guide and a supplement to current technical manuals and other VA criteria in the planning of Research Laboratory facilities.
Drug Enforcement Administration (DEA)	Security Outline of the Controlled Substances Act of 1970	Practitioners include physicians, dentists, veterinarians, researchers, hospitals, pharmacies, or other persons registered to do research, to dispense, or to use in teaching or chemical analysis a controlled substance in the course of professional practice. Minimum security standards for practitioners are set forth in the regulations and are to be used in evaluating security.
General Services Administration	P-100 Facilities Standards for the Public Buildings Service	The Facilities Standards for the Public Buildings Service establishes design standards and criteria for new buildings, major and minor alterations, and work in historic structures for the PBS of the GSA. This document contains policy and technical criteria to be used in the programming, design, and documentation of GSA buildings.

Issuing Organization	Title	Description
Leadership in Energy and Environmental Design	What LEED is	LEED, or Leadership in Energy and Environmental Design, provides building owners and operators with a framework for identifying and implementing practical and measurable green building design, construction, operations, and maintenance solutions.
National Fire Protection Association	NFPA 45: Standard on Fire Protection for Laboratories Using Chemicals	Essential to the fire-safe design and operation of instructional, educational, and industrial laboratories using chemicals, NFPA 45: Standard on Fire Protection for Laboratories Using Chemicals outlines the maximum allowable quantities of liquids and gases, as well as requirements for laboratory ventilating systems and chemical fume hoods.
National Institute for Occupational Safety and Health (NIOSH)	Prevention Through Design	One of the best ways to prevent and control occupational injuries, illnesses, and fatalities is to "design out" or minimize hazards and risks early in the design process. NIOSH is leading a national initiative called Prevention through Design to promote this concept and highlight its importance in all business decisions.
National Institute of Building Sciences	Whole Building Design Guide (WBDG)	Within the WBDG, research facilities are divided into two major groups: animal research facilities and research laboratories. Research laboratories are further categorized by type (e.g., wet labs and dry labs) and by sectors (e.g., academic, corporate, and government labs). This site provides a wealth of information and available resources addressing each laboratory type.
National Institutes of Health	Design Requirements Manual	This manual is the only detailed design requirements and guidance manual for biomedical research laboratory and animal research facilities in the United States. Provides minimum performance design standards for NIH-owned and -leased new buildings and renovated facilities, ensuring that those facilities will be of the highest quality to support biomedical research.
Office of the President	EO13514 Federal Leadership in Environmental, Energy, and Economic Performance	EO 13514 sets sustainability goals for Federal agencies and focuses on making improvements in their environmental, energy, and economic performance. The executive order requires Federal agencies to set a 2020 greenhouse gas emissions reduction target within 90 days, to increase energy efficiency, to reduce fleet petroleum consumption, to conserve water, to reduce waste, to support sustainable communities, and to leverage Federal purchasing power to promote environmentally responsible products and technologies.
Office of the President	EO 13423 Federal Leadership in Environmental, Energy, and Transportation Management	EO 13423 requires agencies to reduce greenhouse gases through a reduction in energy intensity of 3% a year or 30% by the end of fiscal year 2015. Federal agencies must ensure that at least half of renewable energy comes from new renewable sources and agencies must reduce water consumption by 2% annually through fiscal year 2015. In addition, new construction/major renovations must comply with the 2006 Federal Leadership in High Performance and Sustainable Buildings Memorandum of Understanding.

Issuing Organization	Title	Description
U.S. Department of Energy	Energy Star Rating System	The national energy performance rating is a type of external benchmark that helps energy managers assess how efficiently their buildings use energy, relative to similar buildings nationwide. The rating system's 1 to 100 scale allows everyone to quickly understand how a building is performing—a rating of 50 indicates average energy performance; a rating of 75 or better indicates top performance.

CONSTRUCTION STANDARDS

Issuing Organization	Title	Description
International Organization for Standardization	ISO 14644: Cleanrooms and Associated Controlled Environments	This eight-part standard covers the classification of air cleanliness; specifications for testing and monitoring to prove continued compliance; test methods; design, construction, and start-up; operations; vocabulary; separative devices; and classification of airborne molecular contamination.
Occupational Safety and Health Administration	Standards – 29 CFR	This part sets forth the safety and health standards promulgated by the Secretary of Labor under section 107 of the Contract Work Hours and Safety Standards Act.
United States Army Corps of Engineers	Construction Contractor Submittals	The chapter on the submittals process describes the purpose of submittals and responsibilities of various parties to submittal management.
United States Government	Federal Acquisition Regulation	This regulation provides guidance for Federal contracting

RELOCATION STANDARDS

Issuing Organization	Title	Description
National Archives and Records Administration	Code of Federal Regulations, 49 CFR Part 172	This section of the code covers transportation regulations with a focus on hazardous materials, special provisions, hazardous materials communications, emergency response information, training requirements, and security plans.
National Archives and Records Administration	Code of Federal Regulations, 49 CFR Part 173	This section of the code covers transportation regulations with a focus on shippers, general requirements for shipments, and packaging.
National Archives and Records Administration	Code of Federal Regulations, 49 CFR Part 177	This section of the code covers transportation regulations with a focus on carriage by public highway.

Issuing Organization	Title	Description
International Air Transport Authority	Dangerous Goods Regulations	These regulations offer guidance in various formats to ensure correct preparation and documentation before shipping dangerous products.

APPENDIX B: LABORATORY DESIGN CONSIDERATIONS

These guidelines, which can be used to assist in decision making during the Planning and Design Phases, offer details on laboratory equipment and space requirements. The guidelines are intended to provide general context of the variety of design considerations. These considerations will be shaped against the specific requirements defined for each project.

GENERAL LABORATORY DESIGN CONSIDERATIONS

These considerations are general and apply to all sections within a laboratory. These are only considerations and, as such, may or may not have bearing on any individual facility.

Space/Equipment/ System	Design Considerations
Building Exterior	Bullet-resistant exterior cladding Bullet-resistant glazing on ground floor Discrete entry for evidence delivery Designated parking for evidence delivery Secure vehicle impound yard Attack-resistant exterior wall design
Security	Closed-circuit television (CCTV) coverage in parking lots CCTV coverage in secure vehicle impound yard CCTV coverage at all exterior doors Door status monitors for all exterior doors Intrusion detection for all ground floor exterior windows Motion detection in evidence storage spaces Vehicle ram protection at entrances Secured landscape design to prevent concealment of potential intruders Secured HVAC air intakes Established security perimeter to separate public and secure spaces Monitored proximity access system to all secure spaces and all laboratory sections Electronic security systems monitored locally and remotely after hours Programmable security system that controls access to specific rooms or areas
IT Network	Phone system integration (voice over Internet protocol, call manager, direct inward dialing lines) Determine wired vs. wireless access (or a mix) Networked printing/scanning Network solution for building control systems with attention on remote access by contractors
Auxiliary Power	Provision of uninterrupted power supply and/or generator back-up requirements for key equipment (for example, computer network, instrumentation, access control system, evidence and reagent refrigerators, emergency lighting, emergency response functions within the facility)

Space/Equipment/ System	Design Considerations
Vehicle Processing	12-foot-wide by 20-foot-long vehicle bay 8-linear-foot work bench Shop sink Shop compressed air Vehicle tire removal equipment Tools and equipment storage (may be secure storage cabinets in separate room) Ability to cordon off each bay securely, if multiple bays Quantity of bays depending on staff and caseload Cyanoacrylate (superglue) fuming room with exhaust at 6 inches from floor, vapor-proof light fixture, light switch outside room, epoxy-painted gypsum wall board for walls and ceiling Multifunctional space Durable wall and floor construction Support utilities
Fume Hoods	Class A, chemical fume hood with remotely located and acoustically isolated exhaust blower, preferably an energy efficient model Utility needs, such as laboratory compressed air, laboratory vacuum, water, laboratory gases
Biological Hoods	Class II, types A2, B1, or B2, depending on specific application
Bench-Top Work Surfaces	Solid chemical-resistant material, such as epoxy resin, in laboratory space where fume hood or chemical-rated biological hood is installed Stainless steel in spaces where biologically contaminated evidence is to be placed on the work surface 1.5-inch thick laminated maple in physical examination spaces, such as armorer's bench and machine room bench in firearms section and the vehicle processing work bench Chemical-resistant plastic laminate in all other laboratory spaces not identified above Standard plastic laminate in non-laboratory spaces, such as offices, conference rooms, lunch room
Laboratory Sinks	Epoxy resin in solid chemical-resistant countertops Glassware pegboard with drip tray, as required All other locations, stainless steel
Hands-Free Sinks	All clean sinks Other special locations as specified by users
Clean Sink Areas	At the entrance to every laboratory section Only for washing hands upon leaving a laboratory section Near laboratory coat hangers Near cabinetry for storage of soap, paper towels, gloves, and other protective garments, as determined necessary by the users
Chemical Waste Plumbing	All epoxy resin sinks and fume hood cup sinks Dilution system, neutralizing filters, or holding vessels, depending on local authority and plumbing design, if needed At laboratory sinks and fume hoods, cup sinks where chemical spills or disposal occurs

Space/Equipment/ System	Design Considerations
Ante Rooms	Between clean and dirty spaces (for example, between main circulation corridor and entrance to a laboratory section that potentially contains hazardous airborne contaminants)
	Air handled through directional airflow to prevent exfiltration of contaminated air
	All the features of the clean sink areas described above
	Recommended minimum size: 64 square feet
Bulk Chemical and Clandestine Laboratory Evidence Storage Rooms	Exhaust system to expel both heavier- and lighter-than-air vapors
	Explosion-proof electrical fixtures and chemical spill containment
	Verification of room code occupancy type based upon type and amount of chemical to be stored per local, state and national code requirements.
Finishes	Laboratory floors: chemical-resistant sheet rubber or rubber tiles with welded seams and an integral cove base
	Laboratory walls: epoxy-painted concrete masonry units or gypsum wall board, bullet-resistant wall with epoxy coating at firing range
	Laboratory ceilings: non-shedding acoustical tiles, except in post-amplification rooms and wash-down rooms, such as evidence drying rooms, which are epoxy-painted gypsum wallboard
	Non-laboratory spaces: acceptable interior finish standards for offices and non-laboratory support spaces
Laboratory Cabinets	Standard modular fixed or mobile table-based systems suitable for installation in wet chemistry and biological sciences laboratory spaces
	Standard cabinet configuration of 1/4 drawer cabinets, 1/4 door cabinets, and 1/2 knee space unless users identify specific storage needs
	Steel or wood preferred, plastic laminate acceptable
	Maximum use of mobile flexible laboratory casework systems
Files	Generally, one four-drawer filing cabinet, or the equivalent file storage space, for each analyst at the area of the non-laboratory workstation
	Use of high-density mobile storage system to minimize floor space
Special Considerations	Acoustics to contain sound and minimize reverberation in firing range
	Vibration-proof flooring for scanning electron microscopes as prescribed by the microscope manufacturer
	High-strength floor structure for high-density mobile storage systems and bullet recovery tanks in firearms section
	Scanning electron microscopes located away from electromagnetic fields as required by the manufacturer

Appendix B: Laboratory Design Considerations

Space/Equipment/ System	Design Considerations
Laboratory Section Office Spaces	Supervisor's private office or as required by space standards Verification of office sizes with agency office standards Recommended size minimum: 120 square feet
	Analyst's open-office cubicle or as required by space standards for latent print and questioned document analysts Verification of office sizes with agency office standards Recommended size minimum: 64 square feet
	Case review: for laboratory sections with four or more staff members, table and chairs for half of the analysts and book shelving for references and manuals Verification of office sizes with agency office standards Recommended size minimum: 80 square feet

ADMINISTRATION SECTION

These considerations apply to the laboratory's administrative office areas.

Space Type	Design Considerations
Private Office	Private offices for upper management, including, but not limited to, the following: director, deputy director, quality assurance/quality control manager Verification of office sizes with agency office standards
Open Office Area	Administrative assistant, clerical, visitor reception counter
Administration Workroom	Copy machine, facsimile machine, shredder, mail boxes, administrative supplies, case files, business files, etc.
Public Spaces	Reception/waiting area, visitor reception counter, display cabinets
Toilet Room	Separate men's and women's or unisex, accessible to people with disabilities according to the Americans with Disabilities Act
Conference/Training	Spaces for, including, but not limited to, the following: director's conference room, general conference rooms, classrooms, training laboratory, mock crime scene room, blood spatter training and analysis room

CONTROLLED SUBSTANCE SECTION

This section analyzes evidence to identify material secured during drug seizures.

Space Type	Design Considerations
Main Laboratory	15-linear-foot or 90-square-foot, U-shaped bench, per analyst per user preference
	Individual analyst workspace with secure evidence storage cabinet for in-process evidence
	6-linear-foot miscellaneous bench per analyst for fume hood, laboratory sink, bench space for equipment and procedures, as needed
	Localized vibration-isolation benches for analytic balances, as required
	Floor space for floor-mounted equipment, as required
Instrumentation Room	36-inch-deep bench for each instrument, as needed, with adjacent layout bench, computer, and other equipment, as needed
	Access to the rear of each instrument, as needed, for maintenance and utility access
	Instrument maintenance space with fume hood, solvent storage, and laboratory sink in instrumentation rooms with eight or more instruments
	Shelving for storage of references and manuals
	Dry fire-suppression system
	Sufficient data requirements for each instrument and workstation
	Sufficient power requirements for each instrument
	Sufficient heat-exhaust characteristics for each instrument
	Ventilated, acoustical base cabinets for instrument pumps
	Sufficient type and quantity of laboratory gases
	Gas cylinder closet for gas cylinders with manifolds, plumbed through wall to instruments (gas generators for hydrogen, nitrogen, and zero-air may be utilized in lieu of cylinders)
	Separate enclosed room to control noise and heat
Evidence Storage Room	Additional room to in-process storage in workstations, as required
	Full-height wall partitions secured to structure above
	Dual security access devices per DEA requirements
Standards Storage Room	Laboratory refrigerator
	Standards secured to storage units
	Wall partitions secured to structure above
	May be located in evidence storage area

Space Type	Design Considerations
Reagent Preparation	Separate room or part of the main laboratory 4-linear-foot chemical layout bench 4-foot fume hood with chemical storage base cabinets Under-counter glassware washer Double sink Autoclave, if required 3-linear-foot glassware clean bench 3-linear-foot glassware dirty bench Glassware storage Bench space for specific equipment and procedures, as needed Explosion-proof flammable materials storage refrigerator, as needed
Other Considerations	Room for analysis of large bulk drug seizures, as needed Chemical storage room Space for drying freshly harvested marijuana, as needed Botanical enclosure for growing marijuana, as needed

TOXICOLOGY SECTION

This section analyzes biological specimens for the presence of alcohol, drugs, and other toxic compounds. The analysis can aid medical or legal investigation of death, poisoning, and drug use.

Space Type	Design Considerations
Main Laboratory	15-linear-foot or 90-square-foot, U-shaped bench, per analyst per user preference Individual analyst workspace with secure evidence storage cabinet for in-process evidence 6-linear-foot miscellaneous bench per analyst for fume, as needed, laboratory sink, bench space for equipment and procedures, as needed Laboratory equipment bench space, as needed

Space Type	Design Considerations
Instrumentation Room	36-inch-deep bench for each instrument, as needed, with adjacent layout bench, computer, and other equipment, as needed
	Access to the rear of each instrument, as needed, for maintenance and utility access
	Instrument maintenance space with fume hood, solvent storage, and sink in instrumentation rooms with eight or more instruments
	Shelving for storage of references and manuals
	Dry fire-suppression system
	Sufficient data requirements for each instrument and workstation
	Sufficient power requirements for each instrument
	Sufficient heat-exhaust characteristics for each instrument
	Ventilated, acoustical base cabinets for instrument pumps
	Sufficient type and quantity of laboratory gases
	Gas cylinder closet in corridor outside instrumentation room for gas cylinders with manifolds, plumbed through wall to instruments (gas generators for hydrogen, nitrogen, and zero-air may be provided in lieu of cylinders)
Refrigerated Evidence Storage	Additional room to in-process evidence storage in workstations, as required
	Secured refrigerators in main laboratory or in separate room
	Walk-in refrigerator and/or freezer, depending on quantity of evidence
	Temperature monitoring and recording system
Reagent Preparation	Separate room or part of the main laboratory space
	4-linear-foot chemical layout bench
	4-foot fume hood with chemical storage base cabinets
	Under-counter glassware washer
	Double sink
	Autoclave, if required
	3-linear-foot glassware clean bench
	3-linear-foot glassware dirty bench
	Glassware storage
	Bench space for specific equipment and procedures, as needed
	Flammable materials storage refrigerator

FIREARMS/TOOLMARKS SECTION

This section performs the examination of firearms, ammunition components, toolmarks, gunshot residue on victim clothing, bullet trajectories, and closely related physical evidence.

Space Type	Design Considerations
Main Laboratory	15-linear-foot , 20-linear-foot if comparison microscope is included, or 90-square-foot U-shaped bench, per analyst per user preference
	Individual analyst workspace with secure evidence storage cabinet for in-process evidence, laboratory sink
	6 linear feet of miscellaneous bench space per analyst as required for specific equipment and procedures with map drawer base cabinets for bullet standards
	Floor space for floor-mounted equipment, as required
Comparison Microscopy Room	5 linear feet per microscope
	Light-level control
	Individual microscopes in individual workstations or separate microscopy room, per user preference
Firing Range	Ventilation for airborne unspent powder and lead/metal particulates (see Occupational Safety and Health Administration and National Institute of Occupational Safety and Health regulations)
	Acoustical design for noise containment and minimizing reverberation
	8-linear-foot weapon layout bench
	8-foot minimum head clearance
	Bulletproof enclosure consisting of 8-inch reinforced concrete walls, ceiling, and floor, such as 8-inch concrete masonry units, fully grouted
	Thicker bulletproof enclosure if firing high-caliber weapons
	Suspended bullet and fragment deflection system
	Deflection protection shield in any protruding devices located down range
	Bullet collection trap at end of range
	Consider lead-contaminated material removal through building
	Dedicated air handling systems with on-demand controls
	Ceiling Baffle System for safe weapon discharge
	Target track system where needed
	Ballistic backstop appropriate to range length and weapon catchment requirements
	Abrasion-resistant steel plate at end of range, minimum behind backstop
	Recommended minimum size: Width: 8 feet to 20 feet, Length: 40 feet to 75 feet
Bullet Recovery Room	Ventilated water tank
	Bulletproof enclosure and acoustical treatment, same as firing range
	Clear pathway for oversized water tank installation

Space Type	Design Considerations
Toolmarks Machine Shop (may double as Armorer's Room)	Tools storage Shop sink Drill press Grinder Compressed air Additional bench space as required for tools and equipment
Integrated Ballistic Identification System (IBIS) Room	IBIS workstations Server associated with automation platform, if applicable Dedicated phone line and additional data ports, as required (requirements per the Federal Bureau of Investigation)
Chemical Exam Room	6-foot fume hood with chemical storage base cabinets Laboratory sink Bench space as required for evidence layout and equipment
Ammunition Storage Room	Ante room between main laboratory and bulletproof enclosure (firing range and bullet recovery room) Storage shelving with extra-high weight-bearing capacity for test fire ammunition
Firearms References Room	High-density mobile storage system for the storage of handguns and long guns in the quantity and percentage appropriate to the jurisdiction served
Evidence Storage Room	Additional room to in-process storage in workstations, as required Full-height wall partitions secured to structure above

TRACE ANALYSIS SECTION

This section identifies and compares specific types of trace materials that could be transferred during the commission of a violent crime. These trace materials include human hair, animal hair, textile fibers and fabric, rope, feathers, wood, soil, glass, and building materials. The physical contact between a suspect and a victim can result in the transfer of trace materials.

Space Type	Design Considerations
Main Laboratory	15-linear-foot or 90-square-foot U-shaped bench, per analyst per user preference Individual analyst workspace with secure evidence storage cabinet for in-process evidence, fume hood as needed 6 linear feet of miscellaneous bench space per analyst for fume hood or exhaust snorkel as needed, laboratory sink, bench space for equipment and procedures, as needed Floor space for floor-mounted equipment, as required

Space Type	Design Considerations
Instrumentation Room	36-inch-deep bench for each instrument, as needed, with adjacent layout bench, computer, and other equipment, as needed.
	Access to the rear of each instrument, as needed, for maintenance and utility access
	Instrument maintenance space with fume hood, solvent storage, and sink in instrumentation room with eight or more instruments
	Shelving for storage of references and manuals
	Dry fire-suppression system
	Sufficient data requirements for each instrument and workstation
	Sufficient power requirements for each instrument
	Sufficient heat-exhaust characteristics for each instrument
	Ventilated, acoustical base cabinets for instrument pumps
	Sufficient type and quantity of laboratory gases
	Gas cylinder closet in corridor outside instrumentation room for gas cylinders with manifolds, plumed through wall to instruments (gas generators for hydrogen, nitrogen, and zero-air may be provided in lieu of cylinders)
	Separate enclosed room to control noise and heat
Microscopy Room	4-linear-foot bench for each microscope, except 5-linear-foot bench for comparison microscopes
	Light level control
	Vibration-sensitive room
Scanning Electron Microscope (SEM) Room	8-linear-foot bench
	Room sized for one SEM with required clearances
	Bench space with sink
	Manuals and reference shelves
	Light level control
	Dry fire-suppression system
	Floor structure designed to support vibration tolerances for specific SEM
	Separate room to house heat producing auxiliary equipment
	Recommended minimum size: 120-square-foot room
Evidence Storage Room	Additional room to in-process storage in workstations, as required
	Flammable storage cabinets for arson evidence
	Full-height wall partitions secured to structure above
Standards Storage	Separate room or part of the main laboratory
	Custom storage cabinet for various types of standards
Arson Investigation Room	Explosion-proof fume hood or snorkel exhaust as needed
	Laboratory sink
	Layout bench
	Fire-protection system with fume hood
	Chemical storage cabinet
	Recommended minimum size: 120 square feet

Appendix B: Laboratory Design Considerations

Space Type	Design Considerations
Evidence Examination Rooms	Minimum two rooms, one large exam room and one small exam room
	Large exam room: one mobile table, 5 feet by 10 feet or two mobile tables, each 5 feet by 5 feet with access on all sides and an 8-foot bench with fume hood and laboratory sink, if required
	Small exam room: one mobile table, 3 feet by 6 feet or two mobile tables, each 3 feet by 3 feet with access on all sides and a 6-foot layout bench
	Sufficient quantity of exam rooms based upon staff and caseload
	Evidence hanging apparatus above exam table, if required
	Light-level control
	Low-flow air movement above or near examination areas
	Multiple lighting levels
	Task lighting for close examination

FORENSIC BIOLOGY/DNA SECTION

This section performs serological and DNA analyses of physiological fluids for identification and individualization. The type of material typically examined includes, but is not limited to, blood, semen, saliva, and dental pulp collected at crime scenes and from articles of physical evidence.

Space Type	Design Considerations
Main Laboratory	15-linear-foot or 90-square-foot, U-shaped bench, per analyst per user preference
	Individual analyst workspace with secure evidence storage cabinet for in-process evidence
	6 linear feet of miscellaneous bench space per analyst for fume hood, biological cabinet, exhaust snorkel, laboratory sink, other equipment, and procedures, as needed
	Floor space for floor-mounted equipment, as required
Evidence Examination Rooms	Minimum two rooms, one large exam room and one small exam room
	Large exam room: one mobile table, 5 feet by 10 feet or two mobile tables, each 5 feet by 5 feet with access on all sides and an 8-foot bench with fume hood and laboratory sink, if required
	Small exam room: one mobile table, 3 feet by 6 feet or two mobile tables, each 3 feet by 3 feet with access on all sides and a 6-foot layout bench
	Sufficient quantity of exam rooms based upon staff and caseload
	Evidence hanging apparatus above exam table, if required
	Light-level control
	Low-flow air movement above or near examination areas
	Multiple lighting levels
	Task lighting for close examination
Evidence Storage Room	Additional room to in-process storage in workstations, as required
	Full-height wall partitions secured to structure above
Refrigerated Evidence Storage	Refrigerator/freezers in main laboratory or walk-in cooler/freezers, as required
	Temperature monitoring and recording system

Space Type	Design Considerations
Reagent Preparation dedicated to Forensic Biology/DNA Section	Separate room or part of the main laboratory 4-linear-foot chemical layout bench 4-foot fume hood with chemical storage base cabinets Under-counter glassware washer Double sink Autoclave, if required 3-linear-foot glassware clean bench 3-linear-foot glassware dirty bench Glassware storage Bench space for specific equipment and procedures, as needed Flammable materials storage refrigerator
Pre-Amplification Room	Polymerase Chain Reaction workstations Laboratory refrigerator/freezer Layout bench with laboratory sink Sufficient quantities of workstations based upon staff and caseload
Post-Amplification Room	Entrance through bio-vestibule designed to prevent infiltration and exfiltration Sufficient 3-foot deep benches with rear access for DNA sequencers Sufficient electrical requirements for DNA sequencers Sufficient heat-exhaust characteristics of DNA sequencers Bench space for sufficient thermal cyclers (may be stacked) Laboratory refrigerator/freezer, as needed Fume hood or biological cabinet, as needed Miscellaneous bench space for equipment and layout or robotics bench Exit through bio-vestibule Janitor's closet dedicated to post-amplification room One-way directionality of room design sequence to prevent cross-contamination Ventilated, interlocked pass-through chamber from pre- to post-amplification rooms
Robotics Room	3-foot deep bench with access on all sides
DNA Research Laboratory	Miscellaneous bench space with hood(s) and sink(s), as required for specific equipment and procedures Recommended Minimum Size: 200 square feet, minimum
Combined DNA Index System (CODIS) Workstations	CODIS computer terminal in each analyst's office workstation or at a centralized location

LATENT PRINTS SECTION

This section performs examinations and comparison of latent fingerprints, palm prints, and footprints.

Space Type	Design Considerations
Main Laboratory	15-linear-foot or 90-square-foot, U-shaped bench, per analyst per user preference Individual analyst workspace with secure evidence storage cabinet for in-process evidence 6 linear feet of miscellaneous bench space per analyst for layout of large cases and as required for specific equipment and procedures Floor space for floor-mounted equipment, as required
Evidence Storage Room	Additional room to in-processing storage in workstations, as required Full-height wall partitions secured to structure above Large floor space for bulky items
Cyanoacrylate (Superglue) Processing	Separate room or part of the main laboratory Sufficient multi-chambered cabinets with individually controlled exhaust
Dusting Stations	5 linear feet per station Bench space for dusting station Separate room or part of the main laboratory Designed for re-circulating models Task lighting Sufficient quantity of dusting stations based upon staff and caseload
Chemical Processing	Separate room or part of the main laboratory Bench space with fume hood(s) and laboratory sink(s) as required for specific equipment and procedures Humidity test chamber
Alternate Light Source Room	10-liner-foot bench for photocopy stand and layout Light level control Recommended minimum size: 80 square feet
Automated Fingerprint Identification System (AFIS) Workstation	5 linear feet per station AFIS computer terminal(s) located in an office environment consistent with the laboratory's planned work flow process

QUESTIONED DOCUMENTS SECTION

This section analyzes evidence appearing on paper. Analysis can include review of handwriting, hand printing, typewriting, printing, erasures, alterations, indented writing, and obliterations.

Space Type	Design Considerations
Main Questioned Documents Laboratory Space	15-linear-foot bench or 90-square-foot, U-shaped, per analyst per user preference Individual analyst workspace with secure evidence storage cabinet Northern exposure with natural light Miscellaneous bench and floor space, as required, for specific equipment and procedures and laboratory sink Two mobile tables, each 5 linear feet by 5 linear feet, for case examination Approximately one table per examiner
Evidence Storage Room	Additional room to in-processing storage in workstations, as required Full-height wall partitions secured to structure above
Instrumentation Room	Bench space for instruments and procedures, as required, and a laboratory sink

DIGITAL FORENSICS SECTION

This section focuses on the recovery and investigation of material found in digital devices, often in relation to computer crime.

Space Type	Design Considerations
Main Laboratory	120-square-foot U-shaped workstation per analyst Individual analyst workstations with secure evidence storage cabinet for in-process evidence Miscellaneous bench for equipment and procedures, as required, with laboratory sink Multiple electrical and data outlets at all workstations Black-out window blinds at all windows (interior and exterior) because of investigation of sensitive materials
Evidence Storage Room	Additional room to in-process storage in workstation, as required Heavy-duty shelving for heavy/bulky computer components
Digital Video Analysis Room	Light level control Quantity of rooms based upon staff and caseload Recommended minimum size: 100 square feet
Digital Audio Analysis Room	Acoustically isolated room Sufficient quantity of rooms based upon staff and caseload
Radio Frequency Shielded Room	8-linear-foot examination bench per analyst Shielded, interlocked ante room at entry Recommended minimum size: 100 square feet

Space Type	Design Considerations
General Considerations	Electronics workshop for repair, maintenance, and calibration of electronics equipment
	Not a wet chemistry or biological science laboratory
	Casework and workstations with electrostatic discharge protection

PHOTOGRAPHY SECTION

This section supports the forensic science departments by documenting evidence and crime scenes. The photography section can also provide displays and graphics for courtroom use.

Space Type	Design Considerations
Photographer Workstations	Individual photographer workstations
	Miscellaneous bench for equipment, photo finishing, art supplies for court displays, laboratory sink
	Recommended minimum size: 64-square-foot open office workstation per analyst
Equipment Storage Room	Miscellaneous storage for photography supplies, studio lights, etc.
	Recommended minimum size: 100 square feet
Photo Studio	Overhead bridge-and-beam studio lighting system
	Studio backdrop
	Bench for photocopy stands
	3 linear feet by 6 linear feet mobile table
	Recommended minimum size: 400 square feet
General Considerations	Not a wet chemistry or biological science laboratory

EVIDENCE MANAGEMENT SECTION

This section is the central point of evidence receiving and management.

Space Type	Design Considerations
Evidence Receiving	Transaction counter for evidence delivery with evidence clerks on opposite (secure) side of counter to receive evidence
	Evidence forms counter
	Evidence packaging bench
	Packaging supplies
	Waiting area
	After-hours pass-through evidence drop-off lockers (various sizes, including refrigerator and freezer lockers) in wall with drop-off area on one side and secure area accessible to evidence clerks on opposite side
	Electronic door security controlled by staff
	Bullet-resistant transaction wall if necessary for security

Space Type	Design Considerations
Evidence Outgoing	Workspace for processing and packaging outgoing evidence Multiple storage cubicles for outgoing evidence and packing supplies
Open Office Area	Evidence clerk open office workstations Miscellaneous bench for printers, copier, etc. Recommended minimum size: 64 square feet
Evidence Storage Area	Room temperature evidence storage, such as open shelving or high-density mobile storage systems Drug storage with walls secured to structure above and dual security access or other access in compliance with DEA requirements for narcotics storage Valuables evidence storage with higher security space for storage of cash, jewelry, and weapons Refrigerated evidence storage, such as a walk-in cooler Frozen evidence storage, such as a walk-in freezer Evidence drying cabinets, such as re-circulating, non-exhaust models Evidence drying room that includes a slope to floor drain, epoxy floor/wall system, hose bib, and stainless steel hanger rod mounted at 6 feet above floor Multiple evidence storage rooms within evidence storage area Electronic security access control at all evidence storage rooms
General Considerations	Sufficient quantity of all evidence storage spaces based upon caseload

Table C-1 outlines the most popular contract types and provides a brief description of each one's characteristics. Review local jurisdiction regulations for guidance as some locales will not allow all methods.

Table C-1: Popular Contract Types

Contract Type	Description
Design-Bid-Build (DBB)	DBB is the "traditional" project delivery method because it is the most well-known and frequently used in the construction industry and has been the prominent method for public work in the 20th century. DBB's main feature is a separation of contracts, where the user group holds separate contracts with the A/E firm and the general contractor.
Design-Build (DB)	The defining characteristic of DB is that the user group contracts with one entity for the design and construction of a project, which results in a single point of accountability. The designer-builder is usually created as a partnership between an A/E and general contractor firm. The user group holds one contract with the design-builder.
Construction Manager at Risk (CMc)/Early Contractor Involvement (ECI)	CMc, also known as ECI, is a project-delivery method that incorporates beneficial elements of both DBB and DB. Although the user group using CMc or ECI still holds separate contracts with the general contractor (in this context, referred to as "CM") and the designer, the CM is hired during the design process and assists in several areas collectively referred to as "preconstruction services." What makes the CM in this scenario considered to be "at risk" is that the CM holds contracts with subcontractors and takes the construction performance risk.
Integrated Project Delivery (IPD)	IPD is a project-delivery approach that integrates people, systems, business structures, and practices into a process that collaboratively harnesses the talents and insights of all participants to optimize project results, to increase value to the user group, to reduce waste, and to maximize efficiency through all phases of design, fabrication, and construction.
Bridging Architect	DB with bridging is a delivery method that combines elements of DB and DBB in which the user group hires and works with an architect to achieve a preliminary design concept and then incorporates this design into the DB procurement that the design-builder uses to complete the construction documents. The "preliminary design" architect then usually serves as the owner's representative during construction.
Public–Private Partnership (PPP)	PPP is a contractual agreement between a Government agency and a private business, in which the private party provides a public service or public project and assumes substantial financial, technical, and operational risk in the project. In return, the private party leases, rents, and/or buys back the facility over a certain number of years.

The most suitable contract configuration for the project depends on the laboratory's resources and policies. Once a contract type is selected, the user group must decide on the method of procurement. This can be done via a sealed bid, quotations, sole-source selection, a sealed proposal, or various other methods. Most major construction projects are procured through a competitive procurement, requiring project owners or their agents to select a method for reviewing and selecting the winning contractor. Although there are various selection methods, the two primary methods used in construction are outlined in table C-2.

Table C-2: Primary Contractor Selection Methods

Selection Method	Description
Low Price Technically Acceptable (LPTA)	The LPTA process is appropriate when best value is expected to result from selection of a technically acceptable proposal with the lowest evaluated price. FAR 15.101-2(a) outlines the best value "continuum," which considers that the relative importance of cost or price will vary. If requirements are clearly definable and the risk of unsuccessful performance is minimal, cost or price may properly become the predominant source selection discriminator; therefore, LPTA may be appropriate.
Best Value	A Best Value selection is a selection process for construction services in which total construction cost as well as other non-cost factors (e.g., technical excellence, past performance, and personnel qualifications) are considered in the evaluation, selection, and final award of construction contracts. This selection method is most appropriate when the requirement is less defined, more development work is required, or the performance risk is greater.

The user group determines which procurement option is in the laboratory's best interest during the Construction Phase's initiate stage. The decision may be made as early as the Planning Phase. Guidance and advice on selecting the best bidding approach can be secured from the contracting officer and the A/E team.

Because of the complexity associated with highly technical facilities, such as forensic science laboratories, the use of the LPTA selection method is highly discouraged. The complex and specialized systems used throughout forensic science laboratories requires contractors that are highly experienced in designing and installing these systems to ensure they work as intended. Therefore, non-cost factors, such as past performance (in designing and constructing forensic science facilities) and personnel qualifications (in planning, designing, and constructing forensic science facilities), are far more important considerations than price.

The cost of constructing or renovating a forensic science laboratory extends well beyond the initial capital costs of building the facility. In any major capital construction project, the life cycle costs of the facility should be factored into the Construction Phase to ensure that decision makers see the complete budgetary picture of a new facility. Costs for a forensic science laboratory can be divided into first costs, which include some user group's costs, and operating costs. Table D-1 outlines the various cost components that laboratory directors should be cognizant of when initially planning a new facility or major renovation. Every project is unique, so costs incurred vary greatly by project; however, the full breadth of potential cost considerations are provided below.

Table D-1: Laboratory Cost Components

First Cost Considerations	Operating Cost Considerations
Forensic science consultant planning services	Routine maintenance
Owner's agent services	Preventative maintenance
Design/engineering services	Emergency repair
Construction services (general contractor plus markups)	Capital improvements/renewal
Construction management services (owner's agent)	Service contracts (e.g., janitorial, trash, shredding, landscaping, or copier service)
Commissioning agent services	Technology renewal/upgrades
Security escorts (owner's cost)	Utility costs
Location/geographic adjustment factors (e.g., labor and materials)	
Operational adjustment factors (e.g., 24-hour operations)	
Specialized relocation costs (e.g., samples and evidence relocation, recalibration of equipment)	
Permitting/registration/certification costs	
Cost escalation	
Project contingencies	

Not all projects require all services or adjustments, so project user groups must carefully review their own organization and its resources to determine which costs can be expected within the individual project.

The Needs Assessment defines the high-level requirements for a project. It is used by user group leadership to define the solution to a level of detail to identify key resources required, such as funding, real estate, and scheduling. With this information, the user group can work with its organization to secure the resources to implement the project. The Needs Assessment seeks to answer the following questions: Why? How big? How much? Where?

This section describes how the Needs Assessment answers these questions, defines the differences between a Needs Assessment and a Design Program, and provides an outline of a typical Needs Assessment.

Why? The project team analyzes the existing building to document its deficiencies and to provide evidence that a new building is needed for the crime laboratory to function in a safe, secure, and efficient manner. This involves a close analysis of overcrowded conditions, potential safety issues, building code violations, and an assessment of the building's mechanical, plumbing, and electrical systems.

How big? To answer this question, the needs assessment consultant and the crime laboratory staff engage in extensive workshop sessions to comprehensively identify every item that occupies space in the new building. This includes, but is not limited to, staff workspaces, equipment, storage spaces, utility spaces, building amenities, conference rooms, training spaces, and shipping and receiving areas. Once this space-consuming information is compiled, it will be thoroughly documented in explicit detail in a spreadsheet format.

How much? Once the consultant team receives the above information, it creates a statement of probable cost for developing and constructing the new building. Because this effort takes place without a Design Program or any building design drawings, this cost estimate is very conservative and is used for budgeting purposes based on cost/square footage.

Where? If a building site has not yet been identified, answers to the above three questions will be sufficient to prepare a reliable needs assessment. However, if a building site has already been selected or if more than one potential site is being considered, the needs assessment will go one step further to include a site evaluation analysis. The site is evaluated in terms of accessibility, topography, location, and other factors. The advantage of having a building site identified prior to executing a needs assessment is the ability to provide more accuracy to the cost estimate.

Typically, the next major task in planning a new facility is preparing the Design Program. This document provides the architectural and engineering design team with all of the information necessary to design the building. It is, in effect, a "recipe" for designing the building.

Whereas the Needs Assessment defines the project to a level necessary to identify the major resources it will require, a Design Program identifies the detailed physical requirements necessary to design the facility. The Design Program encompasses the Needs Assessment while taking information to a much greater level of detail. For example, a Needs Assessment space analysis identifies the square feet and staffing required for each laboratory section. A Design Program defines every room required for each laboratory section. Room definition includes requirements for security, utilities, exhaust hoods, furnishings, environment, power, lighting, communications and data, chemical usage, equipment features, and interior finishes.

Table E-1: Key Elements of a Needs Assessment Report

Section	Content	Details/Considerations
I. Preface	Introductory statements	Statement of authorization to perform the study
	Identification of the team	Architect, engineers, consultants, etc.
	Acknowledgments	
II. Executive Summary	Introduction of the project	• What the project consists of • Summary of legal issues • Accreditation requirements • National technical working groups, e.g., TWG on DNA Analysis Methods, TWG on Materials
	Description of existing facilities	Overcrowding, safety and security concerns, inability to meet codes, recent changes in technology, crime trends, liabilities, accreditation issues, etc.
	Mission statement	• Investigations assistance • Court testimony • Chain-of-custody requirements • Difficulty or impossibility of accomplishing mission under existing conditions
	Summary of all conclusions	• Area • Staff • Site • Budget
	The "bottom line"	All the information necessary to make informed decisions, provided to decision makers
III. Objectives and Methodology	Statement of goals and objectives of the study	• Needs assessment • Design Program • Definition of both
	Identification of methodology	All tasks accomplished to create the document
	Three-step process	Description of process in detail
IV. Trends and Influences (Demographics)	Analysis of emerging social, economic, political, population, environmental, and crime trends	• Description of purpose • Description of methodology
	Relation of data to staff and facility needs	• How trends affect individual laboratory sections and impact future requirements • Data related to individual and overall caseload • Data related to facility needs regarding expansion and flexibility
	Projection of future needs	Data related to new size and nature of facility and staff
V. Facility and Space Descriptions	Description of the entire facility in general terms.	As much narrative information as possible to tell the designers what is unique and unusual about the laboratory.
	Explanation of nomenclature and acronyms	
	Organizational chart	Chart as an introduction to descriptions of laboratory spaces
	Description of spaces by laboratory sections	Purpose and function of each laboratory section (provided so that design team will have better understanding)
	Space data sheets	Sometimes used in lieu of space descriptions and equipment data sheets

Section	Content	Details/Considerations
VI. Adjacencies	Internal: Definition of what components/sections of the building need to be placed next to each other	• Evidence flow • Work flow • Promotion of staff interactions/interactions with supervisor(s) • Compartmentalization/public access
	External: Relationship to client, such as medical examiners, law enforcement, district attorney, and courts	
VII. Move-In and Future Staff and Space Needs	Milestones	

APPENDIX F: KEY ACTIVITIES CHECKLISTS

	Process Step	Activity
	Planning Phase	
☐	2	Mission Requirements Identified in **Self Evaluation**
☐	3	Technical Manager Appointed
☐	3	Project Team Assembled
☐	3	**Project Management Plan** Initiated
☐	3	Baseline Project Budget Developed
☐	4	External Consultant Identified
☐	4	Consultant Funding Acquired
☐	4	Consultant SOW/RFP Defined
☐	4	Consultant Contracted
☐	5	Project Team Completed
☐	5	Project Management Plan Completed
☐	6	**Needs Assessment** Prepared
☐	7	**Design Program** Prepared
	Design Phase	
☐	1	External A/E Consultant Identified
☐	1	Funding Approval Secured for A/E Consultant
☐	1	A/E Consultant SOW/RFP Defined
☐	1	A/E Consultant Contracted
☐	2	Design Program Reviewed with A/E
☐	2	Design Program Revised/Updated
☐	3	**Design Charrette** Executed
☐	4	Preliminary Concept Developed
☐	5	**Schematic Design** Selected
☐	6	Major Systems Selected
☐	7	Design Value Engineered
☐	8	**Design Development Drawings** Completed
☐	8	**Project Cost Estimate** Completed
☐	9	**Construction Drawing** Completed
☐	9	**Construction Specification** Completed

		Construction Phase
☐	1	Construction Team Configured
☐	1	Contracting Method Selected
☐	2	Construction Funding Committed
☐	3	Construction Contract Developed
☐	3	**Construction Contract** Awarded
☐	4	Laboratory-Related Submittals Reviewed
☐	4	Construction Process Tracking Planned
☐	4	Material and Systems Submittals Reviewed
☐	5	Facility Commissioning Validated
☐	6	Final Inspection Conducted
☐	6	**Punch List** Completed
☐	6	**As-Built Drawing**, Major System Manuals Received
☐	7	Facility Operations Staff Trained
☐	8	Warranties Evaluated and Documented
		Relocation Phase
☐	1	Move Captain Designated
☐	1	**Integrated Master Schedule** Prepared
☐	1	**Task Tracker Tool** Assembled
☐	2	Contracted Move Service Requirements Identified
☐	2	Contracted Services Acquired
☐	3	**Equipment Inventory** Prepared
☐	4	Move Responsibilities Coordinated
☐	5	All Materials Packed
☐	6	New Site Material Installations Coordinated
☐	7	Old Laboratory Vacated to Manage All Applicable Compliance Issues
☐	8	Equipment in New Location Calibrated/Validated
☐	8	Staff Moved into New Laboratory

Appendix F: Key Activities Checklists

The purpose of these procedures and checklists is to ensure that laboratory staff moves are accomplished safely.

Vacating a Laboratory
Use this detailed checklist to prepare for a move.
Housekeeping
☐ Remove broken glassware and non-contaminated sharps.
☐ Decontaminate all laboratory equipment work surfaces and supplies.
☐ Package all glassware and supplies in boxes. Do not over pack boxes. Keep boxes less than 30 pounds.
☐ Remove all media and supplies from drawers, shelves, and cabinets.
Biohazardous Materials
☐ Disinfect all work surfaces that may be contaminated with biological agents.
☐ Place all sharps in sharps containers and dispose.
☐ Disinfect and dispose all potentially bio hazardous waste.
☐ Have biological safety cabinets professionally decontaminated prior to moving and recertified after the move.
Radioactive Materials
☐ Survey facility and equipment for contamination by counter meter or radioiodine survey meter, as appropriate, and then wipe tests. Place meter printout and corresponding map in the laboratory radiation safety records.
☐ Clean surfaces and equipment if contamination is detected above three times background. If nonremovable contamination is detected, contact the agency, lab, or local authority's radiation safety services.
☐ Schedule radioactive waste pickup services. (Request a waste pick up within 3 days.) Coordinate timing with the end of radioactive material use and before the move date.
Chemical Safety
☐ Remove all laboratory chemicals including waste from laboratory (request a waste pickup).
☐ Remove all empty bottles and cans. Empty all containers, deface the label, and remove the cap removed before placing it into the regular trash.
☐ Remove disposable liners/covers from work surfaces.
☐ Wash laboratory bench tops with soap and water.
☐ Empty the fume hoods and wipe down all inner surfaces.
☐ Run water into all sinks and floor drains to fill traps.
☐ If perchloric acid has been used in the chemical fume hood not specifically designed for that purpose, contact the agency, lab, or local authority's chemical safety services for testing to ensure maintenance worker safety

	Laboratory Closeout and Decommissioning *Use this checklist while packing and preparing for the movers. These activities should take place at least 1 month prior to relocation.*

General

☐ Review any lab decommissioning policies applicable to your lab, agency, and local authority.

☐ Confirm if investigators relocating or vacating facilities are responsible for leaving laboratories in a suitable state for reoccupancy or renovation.

☐ Remove all equipment and supplies from the laboratory unless special arrangements have been made for storage or transfer to another occupant.

☐ Remove broken glassware and noncontaminated sharps from the laboratory in rigid puncture resistant containers.

☐ Never transport hazardous materials in personal vehicles.

☐ Affix packing list to each shipping container.

☐ Contact the moving companies to review their requirements because private moving companies may have additional packing requirements. Contract with movers who have a hazardous materials security plan.

☐ Do not use corridors as holding areas or temporary storage; keep corridors free and clear of obstructions.

Chemicals

☐ Remove all laboratory chemicals and chemical containers including wastes from the laboratory.

☐ Ensure all hazardous waste and unwanted empty chemical containers have been properly discarded. (Laboratory personnel may be responsible for filling out hazardous waste disposal tags and arranging for hazardous waste pick-ups.)

☐ Clean and decontaminate storage cabinet areas, chemical residues, drips, and spills.

☐ Remove bench coats and/or disposable liners/covers from work surfaces.

☐ Wash laboratory bench tops with recommended soap and water.

☐ Remove all debris from the fume hoods. Decontaminate and wipe clean fume hood base surface and walls.

☐ Flush floor drains and sink traps with water to prevent backflow of sewer gas.

☐ If necessary, conduct a test for perchlorates to ensure the safety of maintenance workers if perchloric acid has been evaporated or used in significant quantities without scrubbing or trapping of vapors.

☐ Remove signage for specific hazardous materials that are no longer present (e.g., registered carcinogens, eye protection).

Radioactive Materials

☐ Follow laboratory procedures for close-out survey for all radiation-use areas as required by organization.

☐ Remove radiation warning signs and labels from doors, hoods, and benches.

Biohazardous Materials and Sharps Containers	
☐	Dispose of needles, syringes, and other biohazardous sharps in biohazard sharps containers.
☐	Decontaminate work surfaces that may be contaminated with biological agents.
☐	Remove placards and/or biohazard signs from doors and areas within the lab.
☐	Remove all media and supplies.
☐	Decontaminate biological safety cabinets.
☐	Determine if partially full, red plastic biohazard sharps containers may be capped, taped, and moved to the new location.
Compress Gas Cylinders/Cryogenics Cylinders	
☐	Label each cylinder to identify whom it belongs to and what it contains.
☐	Remove regulators and make sure valve protection caps are securely in place.
☐	Determine who is moving gas cylinders and ensure proper moving equipment (may use gas cylinder carts for transporting).
☐	Determine process for special cylinder relocations needs.
Spills	
☐	If you have a spill involving hazardous chemicals, contact proper authorities or 911 for emergency assistance. DO NOT clean spills without assistance unless you have been trained to do so.
Waste Disposal Checklist	
☐	Find out if a chemical waste pick-up request form for unwanted hazardous materials and hazardous wastes is required and, if so, contact the appropriate party.
☐	Find out if a separate process for discarded hazardous material as a reusable chemical reagent exists.
☐	Identify and label all unknown waste materials and segregate wastes by hazard class.
☐	Ensure that each container is completely leak-proof or sealed (e.g., lids or caps tightened, debris double bagged and air tight, and questionable containers double bagged) for transportation.
☐	Where appropriate, pack chemical wastes in approved shipping boxes and allow enough slack between containers to accommodate handling and transportation? Because some labs may need to inspect box contents before transport, do not seal the boxes.
☐	Clearly mark outdated peroxide formers and other reactives and segregate for separate handling and evaluation.
☐	Label each shipping box with the words "Hazardous Waste" and the specific hazard class of the chemicals.
☐	Store hazardous wastes in an obvious unobstructed area in the lab for pick-up and transport. Check if procedures need be followed to meet licensing requirements.

Move-In Preparation
Use this checklist to ensure the new space is prepared for the movers.

Overview

☐ Minimize the amount of materials to be moved by purging equipment, supplies, and materials.

☐ Designate a single point of contact to manage the move, evaluate equipment, and generate an equipment matrix.

☐ Arrange a detailed pre-move site visit to review equipment sizes, egress path, loading dock, and freight elevator.

☐ Identify where all equipment is going, start-up time, status of current experiments, and when the equipment can be shut down.

☐ Hire a certified chemical mover with knowledge on purging, packing, and transportation licenses.

General Conditions

☐ Complete commissioning of space prior to move-in

☐ Post hazardous work areas and equipment for biohazards, carcinogens, radioactivity/radiation, lasers and UV light, and other potential hazards.

☐ Secure compressed gas cylinders firmly.

☐ Ensure there is a minimum of 36" clearance in the aisles.

☐ Store heavy/hard objects in low, easy reachable locations. Ensure overhead objects are well secured.

☐ Post emergency procedures to be viewable by all lab users.

☐ Identify and post clean areas locations.

☐ Label refrigerators/freezers as "No Food or Drink" or other appropriate labels, such as "No Chemicals," "No Radioactive Materials," and "No Biological Agents," as appropriate.

☐ Secure cabinets and shelves over 48" high firmly to the wall.

Emergency Equipment

☐ Locate emergency eye wash and shower access within 100 feet of lab.

☐ Ensure fire extinguishers are accessible within 50 feet of lab.

Chemical Storage

☐ Segregate chemicals (including waste) by hazard class.

☐ Identify specific chemical incompatibilities.

☐ Store acids away from bases.

☐ Store corrosive materials in low cabinets or shelves below waist height.

☐ Ensure flammables are correctly stored and kept away from oxidizers.

☐	Label different containers for radioactive, chemical, and biohazardous waste and store separately by group.
☐	Verify chemical storage shelves have lips or guards.
	Fume Hoods
☐	Clean and certify fume hoods.
	Biosafety Issues
☐	Label and provide sharps containers for broken glass and needles.
☐	Submit and amend a biological use authorization form (if required).
	Radiation Safety
☐	Post and identify clean areas.
☐	Post "Caution, Radioactive Material" signs on doors to radioisotope labs.
☐	Confirm the radiation use authorization has been amended to include radioisotope use at the new location.
☐	Appropriately label and shield waste storage areas.
☐	Verify all radioisotope work surfaces are covered with plastic-backed absorbent paper labeled with radioactive material caution tape.
☐	Designate all appropriate refrigerators, freezers, fume hoods, and equipment items with "Caution, Radioactive Material" signs or labels.

	Preoccupancy *Immediately after the move, this checklist should be used to warrant that the new laboratory is functioning safely and efficiently for the employees.*
	Emergency Equipment
☐	Ensure emergency eye wash and shower are working and accessible within 100 feet (or 10 seconds) of the lab work areas
☐	Confirm fire extinguishers are accessible within 50 feet.
☐	Ensure appropriate spill kits are available and accessible.
	General Conditions
☐	Acquire any required biological safety cabinet recertification for proper operation.
☐	Provide proper disposal containers for sharp materials (needles, broken glass, etc.).
☐	Verify storage clearance within 18-inches of the ceiling to ensure it will not interfere with fire sprinklers, etc.
☐	Provide a minimum of 24-inch clearance in the aisles (or required clearance by code).
☐	Store heavy items low.
☐	Seismically secure tall pieces of equipment.
☐	Store all toxic gases in a mechanically ventilated area ,such as a toxic gas cabinet or fume hood.
	Chemical Storage
☐	Properly store flammables.
☐	Confirm liquids have secondary containment.
☐	Designate an area for collecting unwanted chemicals for removal.
	Radiation Safety
☐	Ensure materials are properly stored.
☐	Verify radioactive labels are properly posted on cabinets, hoods, and refrigerators where material is located.
☐	Post "Emergency Procedures" signs.
☐	Post any required "Notice to Employees" signs.

www.ingramcontent.com/pod-product-compliance
Lightning Source LLC
Chambersburg PA
CBHW081549170526
45166CB00009B/2629